Polywater

Felix Franks

POLYWATER

The MIT Press
Cambridge, Massachusetts
London, England

This book was set in Univers by The Colonial Cooperative Press, Inc. and printed and bound in the United States of America.

Excerpt from the book *Cat's Cradle* by Kurt Vonnegut, Jr. copyright © 1963 by Kurt Vonnegut, Jr. Reprinted by permission of Delacorte Press / Seymour Lawrence.

Library of Congress Cataloging in Publication Data

Franks, Felix.
    Polywater.

    Bibliography: p.
    Includes index.
    1. Polywater.   I.   Title.
QC145.48.P6F7   546'.22   80-22972
ISBN 0-262-06073-6

Contents

Preface

Friends and acquaintances whose opinions I highly respect advised me that any attempt to write a chronicle of polywater would only lead to frustration and bitterness, because I would disinter much that had best be forgotten. I would be quoted out of context; sociologists with axes to grind would use what I had written as ammunition against scientific research as it is currently practiced by most of us; legislators who hold the purse strings would use such a book as a justification to cut funds to universities and other research institutions. The sum total of all this advice was that an analysis of the polywater affair would be seized upon by various sectional interests united mainly by their dislike and distrust of scientists. This may indeed come to pass, but anyone who wants to beat the scientist can find much better material from which to fashion a stick.

I might still have desisted from writing this book had it not been for advice and encouragement that far outweighed the pessimism. Before deciding whether to go ahead, I contacted many of those who to my knowledge had been actively involved in polywater research, or who had written reviews on the subject, for personal reminiscences and for assessments of the episode now that it was all over. I asked them to analyze their personal involvement and to suggest what—if anything—the scientific community might learn about the general conduct of its affairs. The response to these enquiries was so positive and so helpful that I was persuaded that more good than harm could come from its public exposition. I am deeply grateful to colleagues in many countries who responded so promptly to my appeal for help and who spent much time putting their memories and views on paper.

Before launching into the story, I need to satisfy any questions that may be asked about my own involvement in polywater research and my credentials for attempting to write a balanced account. My overriding scientific interest during the past twenty years has been water—its physics and chemistry, its role in the promotion and sustenance of life, and its remarkable involvement in all manners of natural phenomena. It was only natural that the news of anomalous water was of more than passing interest to me, and I followed the written reports from their beginnings in the early 1960s. Also, through my own work and writings but especially during a year spent in the United States at a time when resources for scientific research still seemed limitless, I had the chance to meet and correspond with fellow scientists who shared my active interest in water. Some of them subsequently became fascinated by polywater and undertook laboratory investigations. Nearly all of them speculated on the underlying causes of polywater's eccentric behavior—after all, were we not dealing with a substance ostensibly described by the formula H_2O?

Remarkable though it may seem, I never engaged in active research on polywater. When anomalous water received its first publicity in the Western world, I had just taken up a post in an industrial research institute, and all my energies were devoted to equipping new laboratories, interviewing prospective colleagues, and getting a new research program under way. By the time I had the opportunity to join in research activities, it was too late; I had been left behind in the rush. I must admit to an intuitive distrust of the early anomalous-water revelations, but that cannot be an excuse for inaction. I am very conscious of the rebuke by Leland Allen (*Washington Post,* September 2, 1973) that most scientists rejected polywater out of hand for the wrong reason: because it was different.

Though I did stand on the sidelines, and though I did not publish any views on the subject, polywater was fiercely debated in my laboratories. Reports of the experiments were circulated as soon as they were published, and we certainly kept abreast of developments. Therefore, having followed all

that has been written about polywater, knowing many of
those who wrote it, and having a feel for the unique nature of
water, I claim to be as well qualified as anyone to write the
story of this scientific controversy that developed in the
1960s and was then resolved in the remarkably short period
of four years. Also, being a practicing scientist and having
spent my working life in the company of other scientists may
help me to approach the subject sympathetically, yet
critically.

There can be little doubt that the popular press's reports on
polywater reached large audiences. The issues that were
raised transcended the actual physical nature of the strange
liquid; they touched on the conduct of the research. Nowadays any reference to polywater is always tinged with ridicule, but ten years ago many competent and experienced
scientists were quite convinced of its reality. I can see no reason why the scientific and sociological issues raised by this
unique episode should be shrouded in secrecy; they have
certainly not been forgotten by the scientific community. A
discussion of these issues will in no way alter the fact that,
whatever the mechanisms are that determine the rate and
quality of scientific progress, the history of the last century
teaches us that they work quite well.

Acknowledgments

This book could not have been written without the generous help of many of those whose names are associated with the keyword ''polywater'' in the scientific literature. I should particularly like to thank the following for responding to my enquiries and sharing with me their reminiscences and the lessons which polywater taught them: Barbara F. Howell, Robert J. Good, Chester T. O'Konski, M. T. Shaw, A. C. Hall, Joel H. Hildebrand, William A. Adams, F. Menes, Algird G. Leiga, Allen P. Minton, Michael Falk, Jack Middlehurst, Sherman W. Rabideau, Sefton D. Hamman, Milton R. Lauver, D. H. Pell, Philip F. Low, John L. Anderson, Pieter Hoekstra, David A. I. Goring, James C. Glass, R. H. Wentorf, J. E. Lane, Arthur Cherkin, Jerry Donohue, Stephen L. Kurtin, Alec D. Bangham, Walter M. Madigosky, Lionel J. Bellamy, Paul Barnes, M. Kutilek, and Ellison Taylor.

My very special thanks go to Leland C. Allen for a long and helpful correspondence and for material from his private files, to John Finney for generously providing invaluable information from the archives of the late Professor Bernal, to Marcel Pierre Gingold for his spontaneous and overwhelming response to my initial enquiry, to Simon Schaffer for making available the results of his sociological researches on polywater, and to Bill Bascom for his wealth of information and for reading critically one of the several draft versions of the final manuscript.

I also owe a debt of gratitude to Joyce Johnson for her help in assembling the mountain of research material and to my wife Hedy for the preparation of the final manuscript.

Polywater

Prologue

"I remember, shortly before Felix died, there was a Marine general who was hounding him to do something about mud."

"What did the general have in mind?"

"The absence of mud. No more mud. In his playful way Felix suggested that there might be a single grain of something that could make infinite expanses of muck, marsh, swamp, creeks, pools, quicksand, and mire as solid as this desk.

There are several ways in which certain liquids can crystallize—can freeze—several ways in which their atoms can stack and lock in an orderly, rigid way. Suppose that the sort of ice we skate upon and put into highballs—what we call ice-I—is only one of several types of ice. Suppose water always froze as ice-I on Earth because it never had a seed to teach it how to form ice-two, ice-three, ice-four . . . ? And suppose that there were one form, which we will call ice-nine, a crystal as hard as this desk—with a melting point of 130°F. And suppose that one Marine had with him a tiny capsule containing a seed of ice-nine, a new way for the atoms of water to stack and lock, to freeze. If that Marine threw that seed into the nearest puddle . . . ?"

"The puddle would freeze?" I guessed.

"And all the muck around the puddle?"

"It would freeze?"

"And all the puddles in the frozen muck?"

"They would freeze?"

"You bet they would, and the United States Marines would rise from the swamp and march on."

"There is such stuff?"

"No, no, no, no . . . if you'd been listening to what I have been trying to tell you about pure research men, you wouldn't ask such a question. Pure research men work on what fascinates them, not on what fascinates other people."

''I keep thinking about the swamp . . .''

''You can stop thinking about it.''

''If the streams flowing through the swamp froze as ice-nine, what about the rivers and lakes the streams fed?''

''They'd freeze. But there is no such thing as ice-nine.''

''And the oceans the frozen rivers fed?''

''They'd freeze, of course. I suppose you're going to rush to market with a sensational story about ice-nine now. I tell you again, it does not exist.''

''And the springs feeding the frozen lakes and streams, and all the water underground feeding the springs?''

''They'd freeze, damn it! But if I had known that you were a member of the yellow press, I wouldn't have wasted a minute with you.''

''And the rain?''

''When it fell, it would freeze into hard little hobnails of ice-nine and that would be the end of the world. And the end of the interview too. Good-bye.''

But was there perhaps such stuff after all? When Kurt Vonnegut wrote his science fiction story *Cat's Cradle* he produced ice-nine from his fertile imagination. Ten years later a discovery by a little-known Russian chemist working in a small provincial technical institute laid the foundation of one of the more curious episodes in modern science. He had prepared a substance which became known first as anomalous water and then as polywater, but in essence resembled ice-nine. The fluid was produced from ordinary water, but it did not freeze or boil like water and it hardly resembled water in any of its properties. It took several years for the discovery of anomalous water to make any impact on Western scientists. For a while it was almost treated as a joke. Eventually, however, scientific activity began—first in Britain, and then in the United States, actively supported by defense interests. The number of publications describing polywater increased rapidly, and conferences were hurriedly organized to discuss its nature and possible applications. The United States was still recovering from the ill-fated ''first in space'' contest, and the Russians were not going to be allowed to win again.

Millions of dollars (and rubles?) were spent in the pursuit of anomalous water, and thousands of pages were filled with polywater stories, scientific and otherwise, over a period of several years. A warning that polywater, like ice-nine, might be the most dangerous material on earth alerted the news media, and from then on much of the scientific debate was carried on in the pages of newspapers. The news media were able to invade the sheltered world of the laboratory and to provide a wide public with a glimpse of how scientists operate, how they are motivated, how they communicate with one another, and how they test discoveries and reach their verdicts.

Eventually polywater was recognized for what it was: a nondiscovery, an artifact. In the history of modern science it achieved a unique position. The episode occurred at a time when scientists enjoyed an enviable position in society, when anything seemed possible. The substance in question—water—was one of universal importance, not some esoteric chemical. As a case study of scientific research the polywater story may not be a typical example, but because it caricatures several facets of science and its practitioners the lessons it might teach the professional and nonprofessional alike are thrown into sharp focus.

One may not doubt that, somehow, good
Shall come of water and of mud.

Rupert Brooke

1

Mysteries of the Tap

As far as the scientific community is concerned, polywater is now a dead issue. Yet its memory lingers on in the public's mind. It often creeps in at question periods following general lectures on more acceptable and conservative topics relating to water. The questioner usually recalls reading a newspaper or magazine article about some "new" form of water that was—depending on the viewpoint of the writer—going to greatly improve or destroy our lives. Whatever happened to this polywater? Was it a hoax? Or is research still going on somewhere, perhaps shrouded by secrecy reflecting the delicacy and deep potential of the discovery? The lecturer shrugs. He is used to the question and no longer embarrassed by it. His replies tend to be humorous; it was, after all, simply a mistake.

Polywater as an episode in the history of science is by no means unique. Many other discoveries have failed the test of searching scrutiny, just as many have been rejected out of hand by local scientific establishments only to reappear later in other places to be hailed as major advances. There was, however, something about the enthusiasm with which this particular artifact was pursued, and denounced, that demands our attention. Clearly there were special factors at work.

One such factor is of course the close relationship of polywater to the stuff that comes out of the tap—ordinary water, one of the most abundant, essential, studied, and understood liquids. Or is it? Actually, from a scientific point of view water is far from ordinary. Essential for the maintenance of life as we know it? Certainly. Abundant in the ecosphere? Hardly. Extensively studied. Probably, in the sense that many measurements of its physical properties have been performed, going back several hundreds of years. Best understood? Not at all.

Physicists claim a much better understanding of esoteric substances like liquid helium or liquid nitrogen than they have of liquid water.

In order to understand the enthusiasm with which many experienced scientists became involved with polywater, we must realize that during the mid-1960s the study of liquid water at a fundamental level was a popular pastime among chemists, despite the view held by most physicists that the problems were too difficult. Although a small minority of scientists had recognized that water has several unique and mysterious features which are implicated in its role as universal solvent on this planet, the prevalent view on the state of our knowledge was different. Clearly it was very much a case of familiarity breeding contempt.

The general attitude was, for me, well exemplified by a chance conversation I had with a chemistry undergraduate whom I met on a train. I was returning home after giving a lecture at one of the largest industrial research laboratories in the U.K.; the student had been there for a job interview. He asked about the subject of my lecture. When I replied that I had talked about the structure of liquid water, he looked at me as though I were a lunatic and in a condescending manner informed me that water was just H_2O, and that anything one needed to know about it could be found on half a page of a standard textbook on inorganic chemistry. The clear message was that I was wasting my life. (I found out later that he did not get the job.) I have long since become reconciled to the fact that people take a complete knowledge of water so much for granted that any admission of an interest in its behavior from a physical, chemical, or biological viewpoint gives rise to astonishment or mild amusement.

The fact is that water offends against nearly all the criteria of normality laid down by physicists and chemists. In the case of the biologist this is not quite so clear, since to the best of our knowledge life and water are so inseparable that any other standard of ''normal'' behavior does not exist. In order to understand the doubly abnormal character of polywater, therefore, it is essential to appreciate some of the eccentric properties of so-called "ordinary" water. Although we rarely give them a second thought, they are instrumental in maintaining conditions on this planet which make it possible for life to exist at all.

One can discuss the properties of water at several levels of complexity. We shall begin by comparing it with "normal" substances. This will lead to a consideration of how the physical properties of water affect our terrestrial environment, how water participates in all the reactions that maintain and propagate life, how it acts as a natural environment for many species, and how its distribution and turnover have influenced economic, social, and even political patterns.

Let us look at some quite basic physical properties. There is, for instance, a rule of thumb that the boiling point of a liquid is related to the size of its constituent molecules. In other words, the smaller the molecules the lower the boiling point. A comparison of H_2O with substances having the same molecular size suggests that water should boil at $-93°C$ and that it should freeze only a few degrees below that temperature. It is also well known that most substances are denser in the solid than in the liquid state, but it is equally well known that ice floats on water—that is, the freezing of water is accompanied by a bulk expansion. Again, every school child learns about the maximum density of water at $4°C$, but what is not known is the mechanism whereby a liquid can contract when it is heated. The ecological consequences of this density maximum are manifest: Freezing of rivers and lakes takes place from the surface downward, thus allowing life below the insulating ice layer to continue undisturbed by severe climatic fluctuations.

Much more significant even than the well-publicized density maximum is the abnormally high specific heat of water. Textbooks state that the specific heat of a substance is the amount of energy required to raise the temperature of one gram of the substance by 1 °C. There is also the rule of thumb that, like the boiling point, the specific heat of a liquid is related to the size of its molecules, and yet another rule states that the specific heat of a solid is higher than that of the same substance in the liquid state. We shall see that for water all these rules, and many more, are overturned. The specific heat of liquid water is 1 calorie per gram for each degree rise in temperature. For alcohol the figure is 0.5 calorie, yet the alcohol molecule is three times larger than that of water. On the other hand, when water freezes, its specific heat drops to half the liquid value. All this means that when energy is supplied to liquid water, only half of it is used to raise the temperature; the remainder is stored away in the bulk of the liquid. The ecological implications are staggering: Warm ocean currents, such as the Gulf Stream, move slowly from a region of tropical climate towards the cold regions of the Arctic and the Antarctic, all the while losing heat to the atmosphere. The scale of this heat loss is not generally realized: Every *hour* the Gulf Stream releases stored-up energy to the air equivalent to that generated by the combustion of some 200 billion tons of coal—about two-thirds of the world's annual coal production. In other words, the high specific heat of water enables the oceans to absorb solar energy and act as vast energy reservoirs. As the water masses move slowly to regions of lower temperature, this energy is gradually liberated in the form of heat, a process which is mainly responsible for ensuring a temperate climate, free from violent fluctuations. This unique feature of water has provided Earth with an environment suitable for the development and maintenance of life.

These few physical properties serve to demonstrate that water is anything but a typical liquid. They could be supplemented, because there is hardly a property of water that could be called normal. The main reason for emphasizing water's melting and boiling points, density, and specific heat is that these were the properties that first convinced Soviet scientists that they were dealing with a very strange new species. The early realization that this new form of water, derived from ordinary water, did not freeze or boil like ordinary water, and had a higher density but no density maximum, prompted the obvious question how a substance chemically identical to H_2O could differ from it so markedly.

The practical consequences of the physical eccentricities of water are interesting in themselves and of paramount importance to many branches of science and technology, but the central role water plays in maintaining life is even more fascinating and has some bearing on the manner in which news of polywater reached a wide public. The appreciation of water as something wonderful and mysterious can be traced back in literature and art to biblical times, and its sound and movement have inspired musical masterpieces. Water was one of the four Aristotelian elements, and the realization that it is not an element in the chemical sense dates only from the end of the eighteenth century. The involvement of water in life processes can be considered at several levels: First, we look at the function played by individual H_2O molecules in the biochemical reactions that govern metabolism and growth. Second, we regard water as the solvent medium that transports various essential substances through the living organism. Finally, we look at water as the natural environment in which many plants and animals exist throughout their lives. Let us be aware also that many of the organisms that have developed the capability to live on dry land still begin life in the aqueous environment of the womb. The ability to survive on dry land has not made them any less dependent on water; indeed, this ability has been paid for by the development of highly refined water-conservation mechanisms.

Water takes part in most of the biochemical reactions associated with metabolism. Consider as an example the way humans process food. We eat carbohydrates to provide us with the energy to carry out all the necessary and desirable physiological functions. The utilization of carbohydrates in energy production can be simply represented in chemical terms by the combustion of glucose in the presence of oxygen, with

water and carbon dioxide liberated as the end products. This
is represented by the chemical formula

$$C_6H_{12}O_6 + 6\,O_2 \quad = 6\,CO_2 \quad + 6\,H_2O.$$

glucose   oxygen   carbon   water
dioxide

The daily water turnover of a normal adult human is 2.5 liters. This amount of water is lost by respiration, perspiration,
and excretion and is partly replenished by food and drink.
The shortfall, 0.3 liter, is replaced through the combustion of
food, according to the above equation. The energy generated
by the production of 0.3 liter of water amounts to 32 million
calories. If this were liberated as heat, it would cause a fatal
26 °C rise in the body temperature. Since animals manage to
keep their body temperatures almost constant, the energy
generated during the combustion of food must be channeled
to other uses. This is achieved by a series of subtle and complex reactions, each regulated by an enzyme. Water plays a
part in all these reactions.

If the carbohydrates and oxygen consumed by animals were
not constantly regenerated, the supplies would soon run out.
The natural balance is maintained by the plants, which operate in the reverse direction: They use solar energy to photo-
synthesize carbohydrates from carbon dioxide and water. In
this way 3×10^{12} liters of water are required by the plant
kingdom every day.

Apart from being one of the essential chemicals in the
process of life, water molecules also fulfill the function of
biochemical ''cement.'' In order to promote the reactions
necessary for growth, metabolism, and other physiological
processes, nature has developed substances known as enzymes. These proteins, manufactured in the living cell, are
giant molecules formed by the end-to-end addition of smaller
molecules, the amino acids. They can be regarded as a pearl
necklace, with each pearl an amino acid. The enzyme activity
of these long molecules depends critically on their three-

dimensional arrangement, that is, the manner in which the chain is folded up. Such a folded-up enzyme molecule is shown in figure 1, which well illustrates the molecular complexity associated with biological function. For each protein there is one particular way in which the molecule must fold to exhibit activity, and in the living organism this folding process takes place spontaneously. While we are still quite ignorant of the forces that promote and stabilize this complex folded shape, we do know that water is required. If the composition of the water phase is not exactly right—if it contains too much acid or not quite enough salt—then the protein loses its natural shape and thus its enzyme activity. The manner in which the water cements the protein in the correct shape is still something of a mystery. What is known is that the water cannot form strong chemical bonds with any of the amino acids, but performs its function by weak forces, which can easily be disrupted at the right time and the right place. This too is one of the main features of an enzyme-promoted reaction. The principle of weak interactions is the basis of many natural processes, and in turn such effects are sensitively attuned to those weak interactions characteristic of water. How easily the natural balance can be upset is illustrated when one tries to replace ordinary water with "heavy water," a substance that occurs naturally in small amounts and is chemically almost indistinguishable from H_2O: Only the very simplest forms of life, such as protozoa, can be persuaded to grow on a diet containing heavy water; in all higher forms of life it acts as a cumulative poison.

In addition to its specific involvement in enzyme activity, water also helps to maintain the integrity of other essential biological structures. For example, it stabilizes the well-known double helix of DNA. It is quite astonishing how many biologists seem to be oblivious of the central role played by water in maintaining the complex and interlocking processes that ensure the correct functioning of living organisms.

Let us now turn briefly from the functions of the discrete H_2O molecule to the role of bulk water within the organism. The

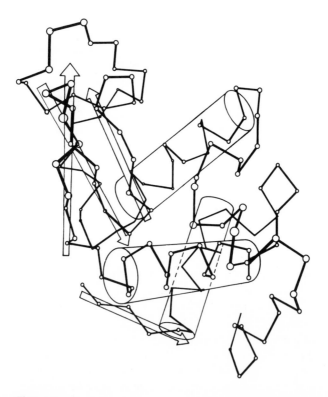

Figure 1
A simplified representation of the folded molecular structure of the
enzyme lysozyme. Each circle represents an amino acid (which itself
contains 10—30 atoms). For the enzyme to be active, it must be in
the precise folded form—the native state—shown here. Water is
known to play an important part in the stabilization of the native
state, but little is yet known about the actual locations of water molecules or the forces involved in such stabilization. The amino acid
regions enclosed in the cylinders and arrows have a high degree of
internal regularity, which is characteristic of many proteins. Such
ordered regions are interspersed with irregular regions that seem to
act as hinges, giving the molecule its complex folded shape in an
aqueous medium. Reproduced with permission from *Characterization of Protein Conformation and Function,* edited by F. Franks (London: Symposium Press, 1979).

fluid carries essential components (such as nutrients, oxygen, hormones, and fats) to sites where they are needed, and also removes waste products for excretion. By means of an example let us return to the 0.3 liter of water produced by the normal human adult from the combustion of food. This reaction requires 185 liters of oxygen, and since air only contains 21% oxygen, which the human lung converts with 14 percent efficiency, this means that 6,300 liters of air must be processed every day. The heart ensures that the oxygen, once it is dissolved in the blood, finds its way to the place where it is required. Each heartbeat pumps 70 cubic centimeters of blood; with 70 beats per minute, the amount of blood pumped around the body each day approaches 7,000 liters. There are also other organs that maintain and redistribute the body's water supply, such as the kidneys, which perform the function of purification. Here again, the quantities of water processed every day (in this case, as urine) are considerable.

The distribution of water throughout the body also requires comment. By weight, over 60 percent of the human body (and 99.5 percent of the jellyfish) is made up of water, but its distribution is by no means uniform. Brain and muscle tissues are richest in water; bone and fat tissues are poorest. Each cell requires a minimum amount of water for correct functioning, but cells can in no way be regarded as bags filled with water. We are quite ignorant about the state and distribution of water within the cell, its function, and what controls its flow into and out of the cell. Of particular relevance to polywater discovery is the phenomenon of ''unfreezable'' water. When biological tissues are cooled, about 20 percent of the tissue water refuses to freeze, even at very low temperatures. The implication is that the water molecules in question are under some kind of constraint that prevents them from taking up their correct places in the ice crystals also present in the tissue. Unfreezable water is not confined to biological materials; it has frequently been observed wherever water exists in confined spaces, such as in the pores of minerals. The existence of unfreezable water does raise the question of how such H_2O molecules, which can only interact with the solid substrate by very weak forces (rather than by strong chemical bonds), can nevertheless be prevented from crystallizing at low temperatures where ice is the only stable form of water. It was this very question—what might be the state of water close to solid surfaces—that led indirectly to the discovery of polywater. Even now that polywater has been shown to be a nondiscovery, the state of water near solid surfaces, and in confined spaces generally, still presents many challenges and unsolved problems.

To complete the short survey of water in biology, we consider how the freakish properties of water have been responsible for the manner in which the bulk liquid can act as a natural environment for plant and animal life. Its high density causes an organism to be 800 times ''lighter'' in water than in air, so that no complicated skeletal structures are required for supporting the weight of an animal or a plant in water. On the other hand, the limited oxygen solubility makes breathing hard work in water. Sophisticated mechanisms for the assimilation of oxygen have therefore had to be developed by plants and animals that normally live in water.

Water and life on this planet are inextricably mixed up at every level, from the H_2O molecule to the bulk fluid, and the development of life as we know it had to be compatible at every stage with the abnormal properties of the only inorganic chemical that occurs naturally in the liquid state on earth. It is this close involvement and compatibility of the inorganic liquid with biological processes that prompted the biologist Andrew Szent-György to describe water as ''the matrix of life.''

A few facts and figures on the distribution and movement of water on and around our earth may help to put in perspective what might have happened if a substance such as polywater could have existed outside the laboratory. Most of the water readily available to us is not fresh, but mineralized. In fact, we live on a knife edge, with only 0.03 percent of the total terrestrial water fresh and accessible. Most of our pure water is locked up permanently in the Antarctic ice cap. Another potential source of fresh water is the water vapor in the atmosphere. Every year 4.5×10^{17} liters of water evaporate. If this were to precipitate uniformly, it would cover the earth to a depth of 106 cm. As it is, 75 percent returns directly to the oceans as rainfall, and most of the remainder returns to the oceans via the rivers or into the atmosphere by evaporation. The total water content of the atmosphere is only 12×10^{15} liters, so the water in the atmosphere is turned over 37 times every year.

The balance sheet of our water supplies shows that over 2,000 years the consumption per person has not changed much from 230 liters per day. The factors that have changed dramatically are the total population on earth and the recent rapid development of water-consuming industries (whose demand increases at the rate of roughly 2 to 3 percent per annum). Since both domestic and industrial waters find their way back into the rivers and oceans, our problems are twofold: to conserve the fresh water supplies now available, and to plan for a gradually increasing level of supply in the future. We have the technical means at our disposal, but past experience suggests that little will be done until the eleventh hour. There is still a widespread belief that public authorities have a duty to supply us with almost unlimited quantities of pure but cheap water. In years to come society will have to learn that water, as a resource, is not something special, but that like other resources it commands a price. Although nature ensures that it is constantly purified and replenished, there are bound to be shortages unless we are willing to pay the going price. It is hard but possible to imagine life without gasoline, but it is impossible to imagine life without water. Ecological, agricultural, climatological, economic, and political aspects of water are currently of great concern to many people and are likely to gain in importance as the years go by. Although these considerations do not figure directly in the polywater story, they provided the general background against which the story developed.

This brief foray into the physical and biological behavior of water and its further ramification into realms far removed from the pure sciences should have dispelled any ideas that water just happens to be one of the many inorganic chemicals on earth, and that it is hardly worthy of detailed study because we know all that needs to be known. The very opposite is the case: At a time when a molecular description of even simple fluids still eludes the physicist, consideration of something as complex as water cannot be contemplated. Chemists, however, must deal more pragmatically with this common and important solvent. It is quite in order to speculate on water's involvement in chemical reactions, even though its fundamental properties cannot yet be accounted for within existing theoretical frameworks. The life scientists are only just beginning to appreciate the subtle ways in which the properties of water dominate and govern metabolism, biosynthesis, and physiology. Any significant progress in our understanding of water's role in single cells, tissues, organs, and organisms will probably have to await advances in the physical sciences.

In 1962, when the first reports of anomalous water became available, we were witnessing a renaissance of research into the physics and chemistry of water and aqueous solutions, but the state of the art was not very well developed. Although there was little in the way of rigorous theory, there was much speculation—some of it very ingenious. It is not surprising, therefore, that the startling findings by Soviet scientists should have caused such excitement and led to such activity among Western scientists. In retrospect, it is surprising how long it took for the wider implications of the Russian experiments to be recognized. It took even longer for the full significance of the polywater hypothesis to sink in after it was first presented to a Western audience in the autumn of 1966.

Let us now retrace our steps and look at the early developments relating to polywater, long before it became known by that name and long before it caught the imagination of the wider scientific community and of the news media.

"Begin at the beginning," the King said gravely, *"and go on till you come to the end: then stop."*

Lewis Carroll, *Alice in Wonderland*

2

Confined Spaces

The experiments that eventually led to the "discovery" of anomalous water were the logical development of research then being pursued vigorously by Soviet scientists into the peculiar behavior of small droplets of water and of water in confined spaces. In order to put the experiments in their true perspective it is necessary to consider the details of this behavior, its manifestations in natural and technological contexts, and the intractable problems raised by the behavior of water generally.

Studies of the peculiar properties of water in confined spaces have a long and distinguished history. Over a century ago Lord Kelvin demonstrated that a very small liquid droplet has a vapor pressure higher than that of the liquid in bulk; in other words, the liquid evaporates more readily from the surface of droplets than it does from a normal large surface. The volatility can be related to the curvature of the droplet, and the effect is such that it only becomes important for drops of micrometer (μm; 1×10^{-6}m) dimensions. The Kelvin equation, which relates volatility and vapor pressure to curvature, has often been used to calculate the properties of water in very narrow capillaries, such as occur in many porous natural materials. The calculations demonstrate that a thread of liquid water confined in a narrow capillary tube should have a vapor pressure appreciably lower than that of water in bulk. Here again, the effect becomes significant only for very narrow pores or capillaries. Capillaries of such small dimensions may be thought to be of little practical importance, but this is far from true. Most natural materials of mineral and of biological origin are constructed of tubes and pores that have just the dimensions at which the Kelvin effect becomes important. Examples are the arteries and veins that are responsible for the circulation of blood in animals and the phloem that fulfills a similar function in plants. Indeed, most animal and plant cells can be regarded as micrometer-dimension droplets. This is even more directly true of the moisture droplets in the atmosphere which we see as cloud formations. The laws governing the nucleation and growth of small water droplets under various atmospheric conditions, which results finally in rain and snow, are still very debatable topics among atmospheric scientists.

The predictions of Lord Kelvin were based on purely thermodynamic reasoning. It was assumed that the vapor pressure of a small drop or a narrow thread of liquid depends only on the curvature and on some of the physical properties of the liquid, and not on any modification of the liquid due to its contact with a solid surface. Nevertheless, over the last sixty years an increasing number of claims have been made to the effect that liquids (especially water) in contact with solid surfaces suffer some influence that causes their vapor pressure to be even lower than predicted by the Kelvin equation. Particularly during the 1930s there were many reports that a liquid in close proximity to a solid surface differs structurally from the bulk liquid. Many of these reports emanated from the Soviet Union, and the name of Boris V. Deryagin was associated with such work right from the beginning.

The word "structure" as applied to a liquid needs some explanation. After all, the essential difference between a solid and a liquid is that molecules in a solid are fixed in certain positions, whereas molecules in a liquid are free to diffuse in a random manner. This is of course true, but if it were possible to take a series of very short-exposure photographs of the liquid the pictures would show that certain geometric arrangements of molecules are favored. For instance, in liquid water any one molecule would more often than not be surrounded by four others placed more or less at the corners of a tetrahedron, as shown in figure 2a. Expressions such as "more or less" or "more often than not" emphasize the essential difference between liquids and solids, for which we could say "exactly" and "always." Translated to the case of

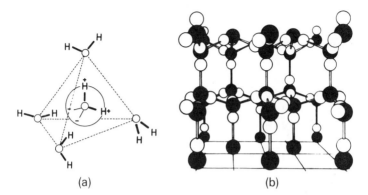

(a)   (b)

Figure 2

(a) Average tetrahedral arrangement and orientation of H_2O molecules in liquid water. This type of arrangement is usually identified with the term "water structure." The central molecule is a more detailed representation of H_2O, showing the distribution of the electronic charge. The molecule can be regarded as a sphere which has embedded in its surface two positive and two negative charges. The positive charges correspond to the positions of the two hydrogen atoms. (b) A permanent molecular arrangement, like that shown in (a), that forms the basis of ice. The black circles represent oxygen atoms and the white circles hydrogen atoms. Because of the regular tetrahedral geometry, the water molecules in ice form an infinite three-dimensional network.

water this means than in ice each H_2O molecule is always surrounded by exactly four others placed at the corners of a regular tetrahedron, as shown in figure 2b. Here the distance between the centers of the molecules is always 0.275 nano-meter (1nm $= 1 \times 10^{-9}$m). One other essential difference between solid and liquid is that successive exposures of the solid would always feature the identical molecules, because they are locked in a crystal structure. In the liquid the mole-cules diffuse, although they do so on a time scale which is very much slower than the exposure time of our hypothetical snapshots. Thus, successive exposures would catch different groups of molecules, but their positions in space would ap-pear to conform to the tetrahedral pattern reminiscent of ice. Therefore, in spite of the violent molecular motions that char-acterize a liquid, individual molecules are ''happiest'' when surrounded by a fairly well-ordered arrangement of other molecules, the actual identities of which change constantly. This arrangement nowadays is given the shorthand descrip-tion ''liquid structure.'' Most of these ideas had not yet been worked out in the days when physicists and chemists first speculated that liquids close to solid surfaces were ''dif-ferent'' from bulk liquids, although the experimental evi-dence seemed compelling. We are now in a better position to enquire into the nature of this difference in terms of liquid structure. The preferred arrangement of molecules in the liq-uid state results from the forces the molecules exert on one another, and it is in this respect that water is unique. The structure of liquid water is therefore quite different from that of more ordinary (in chemical terms) liquids.

Let us pursue the concept of liquid structure a little farther. If the average geometrical arrangement of molecules depends on intermolecular forces, then the presence of a solid surface is likely to affect such arrangements, because the molecules in such a solid are likely to interact with the water molecules in different ways. This is shown diagrammatically in two di-mensions in figure 3. If a collection of small bar magnets, each mounted on a piece of cork, are allowed to float in wa-

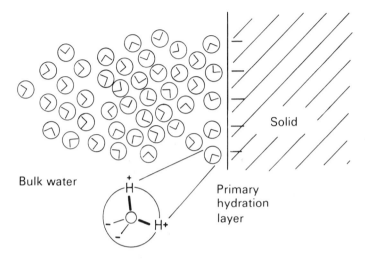

Bulk water

Primary hydration layer

Solid

Figure 3

The effect of a negatively charged solid surface on water structure. Near the surface, water molecules will align themselves with hydrogen (positive) facing the surface (negative). This type of arrangement is not compatible with the normal tetrahedral water geometry that exists far away from the surface. At intermediate distances water molecules are subject to the competing forces due to the surface and to neighboring water molecules. Little is known about the molecular arrangement in this region, or about the thickness of the primary hydration layer. The insert shows a more detailed representation of an H_2O molecule.

ter, the magnets will align themselves so that the positive pole of one will face the negative pole of its neighbor. If we agitate the water, the preferred arrangement of magnets will be disturbed, but unless the agitation is too violent there will still be a residual structure dictated by the magnetic forces. If we then introduce a negatively charged plate into the water, the magnets adjacent to it will reorient so that their negative poles face away from the plate. This will perturb the next layer of magnets, and so on, until at some distance from the plate its influence decays away, helped by the agitation of the water, so that the normal alignment of magnets is once again established. Three regions can be identified: Close to the charged plate the forces due to the plate dominate the arrangement of the magnets. Far away from the plate the magnets are subject only to their own magnetic interactions and do not feel the influence of the plate. In between there is a region where the magnets feel both effects and where their arrangement is completely disorganized.

Such a simple analogy with molecules in a liquid lends a certain attraction to claims that those molecules that sense the influence of a solid surface should be perturbed, and that this perturbation may alter the physical properties of the liquid film close to a solid surface. There is plenty of experimental evidence that this is indeed the case, but a problem arises when we want to quantify the influence of the solid in terms of "liquid-structure modifications." Two questions need to be answered: How is the liquid structure modified from that shown in figure 2a? Over what distances from the solid surface can such a structural modification extend? These questions have not yet been satisfactorily answered, and form the basis of much current scientific debate. At least we now believe that we know the right questions, but this was not the case when the problems of thin liquid films on solid surfaces first received the attention of experimentalists. Probably, the practical implications were realized very early on and provided an additional stimulus to such studies. Nowadays it is generally accepted that the peculiar properties of water in confined spaces contribute to a number of naturally occurring processes, such as the mechanism by which many plants can survive prolonged periods of severe frost. More practical areas where thin liquid films are of significance include the technologies associated with adhesion, drying, and preservation.

In addition to the concept of water structure and its possible modification in the proximity of a solid surface, there is yet another, related phenomenon which forms an important ingredient of the developments that gave rise to the polywater work: the existence of "bound water" (water in an aqueous system that appears to possess properties subtly different from those normally associated with liquid water). The "unfreezable" water referred to earlier is an example of bound water. Implicitly, molecules of bound water are held together by forces other than those normally operating between water molecules. From a practical point of view bound water is important because of its part in determining attributes such as the shelf life of food products and pharmaceuticals, the uptake of moisture from the atmosphere, and the bulk properties of textiles. Many different experimental methods for determining the presence of bound water have been described; the trouble is that different techniques give different estimates of its proportion relative to unperturbed molecules. This is not as serious as it may seem, because there is no sharp dividing line between "bound" and "free" water. What is significant here is the principle that the normal behavior of water molecules in the bulk liquid is easily perturbed by interactions with other types of molecules. Here, as in the case of water structure, the question is how much water can be bound by any other molecule, say, a protein. Opinions differ, as do fashions. Frequently the debate gets to be quite vitriolic, and the scientific community is split into various factions on the subject of "bound water." There are those who believe and claim that all the water inside biological cells—in the cytoplasm—is bound water, whose properties differ substantially from those of the water in the extracellular spaces. This hypothesis is then used to account for many of the processes by which cells are able to regulate the inflow and outflow of all the substances required for their proper functioning. Such ideas are hotly contested by the majority of biologists, but this is not the place to weigh the evidence for or against such proposals. Suffice it to say that concepts such as water structure and bound water are of crucial importance in many different fields of science and technology, which probably accounts for the extensive literature devoted to these subjects. An additional facet, and one that does not make progress any easier, is that the widespread interest in bound water and water structure in its various manifestations has produced many different groups of scientists and technologists, which pursue their studies in complete isolation from one another. Worse still, they operate in ignorance of the existence of other groups with identical interests. Each group publishes in its own journals, organizes its own conferences, develops its own jargon, and has its own ''experts,'' few of whom are aware of similar activities in other groups. Perhaps this state of affairs is not really so strange; who could be expected to guess that cloud physicists and microbiologists might have common problems? The last to realize this would be the cloud physicist and the microbiologist, each enjoying the security of his own group, communicating in his own ''language'' with like-minded people, and meeting with his fellows at regular intervals at international congresses.

Having over the years acquired the label of water expert, I often find myself in the company of such specialist groups as they discuss how ''water problems'' impinge on their particular branches of science. I am constantly reminded how the same problems, only slightly disguised, appear in fields of science and technology that, on the face of it, have nothing in common: atmospheric physics, papermaking, food processing, textile technology, hematology, cell biology, soil chemistry, microbiology, enzymology, glaciology, and others. Though languages and folklore may differ, the problems—conceptual and practical—are identical, or at least closely related. The various groups have much to learn from each other, but this hardly occurs to them, unaware as they are of each other's existence and problems.

The advent of mission-oriented research has done nothing to remove the barriers and demarcation lines that inhibit transference of ideas between disciplines. On the contrary, tunnel vision is the order of the day. In these days of financial stringency it is hard enough for a microbiologist to attend a meeting where he will meet others of a like persuasion. Imagine the reaction of the department chairman or the senior manager if such a microbiologist were to ask for permission and funds to attend a congress of atmospheric scientists. And yet, recent work has shown that certain bacterial spores found on rotting leaves during the autumn provide much better nuclei for the seeding of clouds (rainmaking) than anything previously tried. Here, then, we have the interface between microbiology and atmospheric physics.

Compartmentalization, both voluntary and imposed by financial or managerial pressures, limits horizons and must be partly responsible for the sad fact that much of the published work on the subject of bound water is of indifferent quality; this was true in 1960 and is still true today. In addition, however, the terminology used to describe the observed phenomena in earlier days was not very well developed. It is against this background that we must place the initial investigations that eventually led to the ''discovery'' of anomalous water.

Although by the 1960s Russian scientists were not alone in realizing the importance of thin films of water on a variety of solids, and the influence of solid surfaces on the behavior of liquid films in contact with these surfaces, the study of such phenomena was pursued in any organized manner only in the Soviet Union. One investigator was Nikolai Fedyakin, who at that time worked at the Technological Institute in Kostroma, a medium-sized city 190 miles northeast of Moscow on the upper Volga.

Harrison Salisbury, a distinguished American journalist who has spent half his life since World War II in the Soviet Union, has described the Kostroma region "as backward and isolated a region of Russia as [can] be seen." The city itself is a commercial center with an important linen-milling industry. The Technological Institute, founded in 1932, had at the time three departments: Technology, Forest Engineering, and Engineering. There was also a Textile Research Institute elsewhere in the city.

It can be surmised that in such an environment the research facilities available to Fedyakin must have been fairly limited, especially in comparison with the resources known to have existed at the major research centers in Moscow. Nevertheless, Fedyakin, while studying the behavior of liquids sealed in very narrow glass capillaries, discovered that over a period of days secondary columns of liquid appeared in the top part of the tube, which had previously not contained any liquid. These new liquid columns grew in length to about 1.5 mm, over the period of a month, at the expense of the primary columns. Such a spontaneous separation of an ostensibly pure liquid into two fractions should in itself have been enough to stir the curiosity of perceptive scientists, but even more remarkable was the finding that the liquid that collected in the secondary column was denser than the mother liquid. Fedyakin duly published his results in *Kolloid Zhurnal,* a widely read Russian scientific publication. An English translation became available about a year later. The discussion of the results in the book *Present-Day Conceptions of Bound Water in Rock,* published in 1964 by the Soviet Academy of Sciences, illustrated the connection of Fedyakin's work with the ubiquitous problems of bound water.

It is impossible to know for certain what communications existed between Fedyakin, who was presumably working pretty

much by himself in Kostroma, and members of the powerful
Institute of Surface Chemistry of Moscow. What we do know
for certain is that soon after Fedyakin published his findings
the work was taken over by a powerful group of scientists in
Moscow, and that from then on Fedyakin's name appeared
in print only in conjunction with those of one or more of the
Moscow scientists. This type of takeover of ideas and discoveries is common, and one can but speculate how many of the
well-known phenomena and theories that are linked with a
famous name or that have actually made the name famous
were really discovered by somebody else whose name has
long been forgotten. Certainly, ''modified water,'' as it soon
came to be called, is nowadays associated with the name of
Deryagin, not Fedyakin. I have not been able to find any evidence that Fedyakin ever attended a single conference outside the U.S.S.R. where the subject was discussed. A perusal of scientific publications shows that from 1963 on all
work has apparently originated from Moscow, and the Technological Institute of Kostroma is never mentioned again. Did
Fedyakin move to Moscow to continue his work there? If so,
what became of him? It is not easy to find out, because letters addressed to him are never answered.

The man in the limelight throughout the polywater affair was
Boris V. Deryagin, an eminent and highly respected surface
chemist then director of the Surface Forces Laboratory at the
Institute of Physical Chemistry in Moscow. Born in 1902, he
received his scientific training during the turbulent years following the Russian Revolution and joined the Physical Chemistry Institute of the Soviet Academy of Sciences in 1922.
His studies on the nature of the forces between solids earned
him worldwide recognition, and he is well respected for the
high standards of the experimental work performed in his
laboratory. This has earned him several awards, among them
the M. V. Lomonosov Prize of the Academy and the Order of
the Red Banner of Labor. His insistence on high standards
can make him a severe critic of the efforts of other scientists.
Michael Falk, a Canadian chemist, recalls an informal discus-

sion during which Deryagin was asked whether he really believed that modified water existed. He replied that it was not a matter of belief, it was a matter of performing better experiments. His unyielding attitude is also recalled by Willard Bascom of the Naval Research Laboratory in Washington, D.C., who became involved in polywater some time in 1968 and met with Deryagin in 1970 to tell him of his work—work that cast some doubt on the Russian results. Deryagin struck Bascom as a stern man who seldom tolerated any challenge to his viewpoint: ''He listened, quietly sipping tea through sugar cubes. . . . His response was that we had not done the experiment carefully enough.'' When it was suggested that the scientists at the Naval Research Laboratories had serious doubts about polywater, Deryagin responded that he had been investigating it over the past nine years and currently had twenty-five scientists working on it, and he would not put that much effort into an artifact.

This was the man under whose leadership began the vigorous activities to establish the properties of modified liquids, and modified water in particular. Fedyakin's results had two very important implications. The only way the spontaneous formation of secondary liquid columns could be accounted for was by a lowering of the vapor pressure above the surface of the ''normal'' liquid in the capillary. Such a lowered vapor pressure might then be attributed to structural differences in the anomalous component that separated out and formed the condensate. However, such a hypothesis suggested a very controversial principle, namely a structural memory within the liquid. It was one thing to suggest that a thin film of water influenced by the proximity of a solid surface might be structurally perturbed and modified, but it was much more revolutionary to propose that this perturbation and modification could persist in the liquid once the influence of the solid surface was removed. Yet exactly this seemed to be the case in the column of anomalous liquid observed by Fedyakin. The diameter of the capillary tubes was equivalent in thickness to over 10,000 water molecules. The question was therefore how could the water molecules in the center of the capillary remember the peculiar structure of the solid surface that had given rise to the structurally anomalous organization of the liquid. As we have already seen, the molecules in a liquid are subject to the randomizing influence of diffusion, which would tend to destroy any order imposed on the molecular packing by the capillary surface. For several years before Fedyakin's discovery Deryagin had been involved (he still was in 1979) in debates with Western scientists about the possibility of long-range ordering effects of solid surfaces on liquids, but at no time had such claims for long-range effects implied distances as long as those in Fedyakin's experiments.

However, at the time of the discovery of anomalous water such questions were not yet asked. Instead, a much more basic doubt was expressed: that of chemical purity. Was modified water really just H_2O? It was well known that small amounts of some foreign substances can drastically modify the observed physical behavior of water. For instance, the prolonged exposure of Fedyakin's liquid columns to contact with the glass capillary might have led to the leaching out of various substances from the glass. It was therefore necessary to establish the degree of purity and to use experimental techniques that would minimize contamination. This question of purity and possible contamination persisted through the history of polywater and eventually led to its downfall, but in 1962 this fall was still in the distant future. (To be fair, Deryagin and his colleagues took all possible precautions and performed several tests, enough to satisfy themselves that the observed results were not artifacts produced by impurities.)

Between 1962 and 1966 ten important publications appeared from Deryagin's laboratories, most of them in the *Proceedings* of the Academy of Sciences of the U.S.S.R., of which Deryagin is a corresponding member. As is standard practice among scientists, the first efforts were directed toward repeating Fedyakin's experiments and improving the methods. In particular it was felt to be desirable to reduce the period required for the growth of the modified-water columns. Deryagin, who has a long history of elegant experimental work, soon developed several ingenious techniques for achieving this. Essentially, the method finally adopted consisted of growing the modified-water columns in an atmosphere of unsaturated water vapor in pre-evacuated chambers. Figure 4 shows the main features of the experimental setup developed by Deryagin, which was later also used extensively by Western scientists. Because of the ever-present threat of contamination, quartz capillary tubes were used. Unlike glass, this material is pure and does not contain residues of potentially water-soluble substances. To make doubly sure, the quartz capillary tubes were subjected to further purification treatments. The water used in the experiments was also rigorously purified and carefully checked for the absence of traces of organic residues.

The names that appear on the title pages also bear witness to the number of investigators who must, at one time or another, have been involved in the study of modified liquids. Apart from Fedyakin, who made several more appearances, there were M. V. Talayev, N. V. Churaev, A. V. Novikova, I. G. Yershova, B. V. Zhelezhny, and others. Several years later, in a press interview, Deryagin stated that twenty-five scientists were engaged on anomalous-water investigations at his institute.

The early work can be divided into two stages: Initial studies were directed toward establishing the factors influencing the growth of modified-liquid columns, and later the physical properties of the modified liquids were subjected to detailed scrutiny. As might be expected, the appearance of modified liquids in the capillary depended on the degree of unsaturation of the atmosphere in the vessel, on the temperature, on the diameter of the capillaries (no modified-liquid columns could be grown in capillaries over 100 μm in diameter), and on the presence or absence of air. Most important, liquid columns several millimeters in length could be grown in a mat-

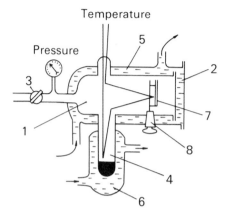

Figure 4

Device used by Deryagin to grow columns of modified liquids in capillary tubes exposed to unsaturated vapor: a cylindrical glass chamber (1) with a flat window (2). Air was pumped out through a valve (3). A tube (4) containing the liquid under test was attached to the chamber, and an unsaturated atmosphere was created in the chamber by suitably adjusting the temperatures of two water jackets (5 and 6). The degree of unsaturation was controlled by adjusting the pressure in vessel 1 and the temperature difference between vessels 1 and 4. The temperature in 6 was always held below that in 5, so that condensation was avoided in those parts of the apparatus not thermostated. The temperature and pressure in the chamber could be recorded. Capillaries (7) for collecting the modified liquids were placed vertically in the middle of the chamber on a stand attached to a plug (8). The appearance and growth of liquid columns was observed by means of a telescope through the window (2).

ter of hours rather than weeks. From footnotes in the published reports of this work in which various well-known scientists are thanked for their help and advice, we can surmise that Deryagin's work created quite a stir in Soviet scientific circles.

The study and documentation of the physical properties of modified liquids grown in the capillaries required the sort of perseverance and experimental ingenuity for which Deryagin was famous. The central problem was one of devising methods for measuring freezing and melting, viscosity, density, and thermal expansion that could be applied to the minute quantities of the modified liquids that were available, preferably without the need to remove them from the capillaries in which they had been grown. Deryagin and his colleagues overcame these problems and developed a set of experimental techniques by means of which they were able to establish beyond any doubt the differences between ordinary and modified water. In the first place, it soon became apparent that the modified liquid that collected in the capillary was a solution of an anomalous or modified component in ordinary water. This posed the further problem of how to isolate the truly modified substance, a problem which was eventually solved. (A short summary of the physical properties of modified water makes it hard for the uninitiated to appreciate the amount of effort that went into the experimental studies and the practical problems that had to be overcome.)

The first finding was that columns of modified water grown by the methods described had viscosities fifteen times the value normally ascribed to water. The thermal expansion, too, showed distinct differences; for instance, in the temperature range 20—40°C it was $1\frac{1}{2}$ times that of ordinary water. The measurements could be exactly reproduced after heating the modified-water column to 150°C, which demonstrated not only that modified water was stable at that temperature but also that it did not boil at 100°C. The cooling of modified water below 0°C also produced striking effects: Solidification occurred only below −30°C, and whatever solid did form was not ordinary ice. The solidification took place over a temperature interval of some $30°C$, quite uncharacteristic of normal freezing behavior. Finally, the density of modified water was found to be $1.1-1.2$ g/cm³ — higher by some 10 to 20 percent than that of ordinary water. The residue remaining after complete removal of the "ordinary" water from the mixture in the capillary had a density of 1.4 g/cm³, and this constituted the "maximally" or "ultimately" modified water. This residue was obtained after subjecting the liquid in the capillaries to prolonged evacuation, and had the appearance of a glassy solid with a boiling point in the region of $250°C$.

These were the basic properties of the mysterious liquid that apparently could be distilled from ordinary H_2O *below its saturation vapor pressure* into narrow capillaries. Following standard practice, Deryagin and his collaborators were more concerned with establishing the reliability of their results than with their explanation; that usually comes later.

During the four years that followed Fedyakin's discovery there was no mention of it in American or British scientific publications, nor did one hear any reactions voiced among British scientists who should have been aware of this work. To be fair, the titles of the Russian reports understated the amazing nature of their contents, and the *Proceedings of the Soviet Academy of Sciences* in the English translation did not have a very wide circulation. Even though few British scientists are able to read and understand Russian, the complete lack of any reaction is surprising. I was at the time dimly aware of Deryagin's activities inasmuch as I dutifully read *Chemical Abstracts,* which provided summaries of the original Russian articles. I must admit that, in common with my fellow scientists, I did not appreciate the importance of the developments taking place.

As is standard practice among scientists, Deryagin reported on his work at several conferences, all of them held in Moscow. In particular, the IUPAC (International Union for Pure and Applied Chemistry) Congress of 1965 took place in Moscow, and here for the first time foreign scientists had a chance to hear about modified water. Dr. Frederick M. Fowkes of Lehigh University was there and recalls that in translation Deryagin's story "was not exciting enough to do more than lift the eyebrows of the skeptics." By all accounts the oral translation system was anything but efficient, and the few American and British scientists who attended the meeting were quite unable to understand the significance of Deryagin's work from the garbled translation.

The first real chance for scientists at large to hear of the new developments came in September 1966, when the Faraday Society held one of its biannual Discussions. The subject was "Colloid Stability in Aqueous and Non-Aqueous Media," and the meeting took place at the University of Nottingham.

The Faraday Discussions enjoy a worldwide reputation for the standard of the contributed papers and for the level of discussion. They live up to their name by being true discussion meetings, unlike most other scientific meetings, which consist of a succession of longer or shorter lectures, with little or no time for questions or discussion because lecturers almost invariably exceed the time allotted to them in the program. Because of their reputation there is often a great competition to have a paper accepted for the Faraday Discussions. The format of the meetings has not changed in many years. A selection committee chooses the twenty papers for discussion. They must all describe original work, and the text must be submitted in full several months before the meeting. Each invited participant receives a set of preprints to read and study prior to the meeting. Authors are allowed only five minutes in which to introduce their precirculated papers, after which they are discussed in full session. The five-minute rule applies to everybody and is strictly enforced by the Secretary by means of colored bulbs: While the green bulb is lit, the contributor is free to speak. At the end of the fourth minute, the green bulb changes for the yellow one; that is the time to become nervous. After five minutes the red bulb lights up,

and it takes a very hardened or famous man to ignore the in-
struction to shut up and sit down. As a result of this strict discipline, everyone gets a chance to have his say and the discussions are not dominated by a handful of people. On the
other hand, the standard and sharpness of the debate can be
such that it requires a good deal of courage—especially for
the younger and less experienced members—to decide to
make an effort to catch the chairman's eye. Everyone who
has contributed verbally to the discussion is given the chance
to put his comments in writing, and even scientists who were
unable to attend the meeting but feel impelled to comment
on any of the papers can do so. The full text, complete with
discussion comments, is published with as little delay as
possible.

The topics for the Faraday Discussions are carefully chosen
for their timeliness, and it is not uncommon for one such
two-day discussion meeting to set the course for the next ten
years' research in a given area of science. Nor is it uncommon to find at such a meeting an international gathering of
famous names in the chosen area. The 1966 Nottingham
meeting provided a good example of the international character of the Discussions—there were participants from the
United States, the Netherlands, Belgium, Yugoslavia, Swit-
zerland, Germany, France, Ecuador, Canada, and South
Africa. Professor Deryagin was the sole representative from
the Soviet Union.

Once again, Deryagin chose an obscure, low-key title for his
paper: "Effects of Lyophile Surfaces on the Properties of
Boundary Liquid Films." This served to disguise, rather than
reveal, the significance of his experimental results. A comparison of the text with Deryagin's past publications (in Russian) shows clearly that the paper submitted to the Faraday
Society was nothing more than a summary of the preceding
three years' work, all of which had already been published in
the Soviet scientific journals. Considering the Society's
jealously enforced rule of originality, this shows how un-

aware Western scientists were of the progress made in Moscow over the past several years.

Contrary to previous practice, however, this time Deryagin allowed himself the luxury of some speculation as to the origin of his observations and their implications. A quotation from his summary will help to illustrate this: "The usual (unmodified) state of water and certain other liquids is thermodynamically metastable"—in other words, modified (or anomalous) water is the *stable* form of water, and it should be only a question of time before all water will turn into the anomalous form, although in practice this may be an extremely slow process. Deryagin again: "For the present discussion it is more essential to emphasize that the experiments described prove the variability of structure of a number of liquids and therefore make it easier to understand their ability to change it under the influence of an interface with a different phase"—that is, the molecules within a liquid can organize themselves in distinctly different ways, and such an organization can originate from contact with certain solid surfaces. Here was something new indeed! It is common knowledge that many chemical compounds, among them H_2O, can exist in different *solid* states (there are eleven different forms of ice, each of them able to exist under certain conditions of pressure and temperature); this phenomenon is known as polymorphism. But to claim such behavior for a liquid that has no permanent, fixed structure was novel, to say the least. It is normally held that the forces holding molecules together in a liquid are much weaker than chemical bonds; Deryagin was now suggesting that a solid surface could alter the forces normally operating between water molecules to such an extent that the modified forces persisted after the effect of the solid surface had been removed. This was to be the explanation for the structural memory, although such a term was never used at the time.

The participants at the Faraday Discussion had had ample opportunity to study Deryagin's paper before the meeting.

One might have justifiably expected that the ensuing discussion would have been an occasion for some very critical and searching questions, and that a lively debate might have followed. This was not to happen. Only ten participants recorded comments on Deryagin's paper. Some dealt with minor technical points, some were irrelevant, and some contributors implied obliquely that they had observed similar phenomena. Professor Dan Eley of Nottingham University and Dr. Karol J. Mysels, both well known for their contributions to the study of liquid surfaces, were the only two whose remarks hit anywhere near the target. Perhaps there were others present who understood the implications of Deryagin's claims but chose to remain silent. This reticence is one of the mysterious aspects of the polywater episode. Was it deference to a famous experimentalist, or courtesy to a well-respected foreigner? It is hard to believe that these were the motives; Faraday Discussions enjoy a ''no holds barred'' reputation. The verdict must be that there was a general lack of perception on the part of the audience.

Despite the seeming innocuousness of this introduction of anomalous water (as it came to be called for the next three years) to a major assembly of Western scientists, seeds of curiosity must have been sown in some minds, because gradually anomalous water became a subject of discussion and speculation among groups of British scientists whose common interest was the study of liquid water. Also, after the Nottingham meeting Deryagin visited several British laboratories and further elaborated on his work. As yet it was purely a matter of listening, asking questions, and weighing the evidence as objectively as possible. What we heard sounded interesting, strange, maybe incredible, but certainly worth following up. So it was that several small groups of scientists in Britain listened carefully to what Deryagin had to tell them and, almost reluctantly, decided to find out for themselves what it was all about.

"Curiouser and curiouser", cried Alice.

Lewis Carroll, *Alice in Wonderland*

3

Rumor and Response

Charting developments up to the end of 1966 has been an easy task, because the only available sources are publications in Russian journals dealing with purely experimental observations. Such personality problems as might have existed even then among Soviet scientists are not common knowledge. Looking back through the reports, one is struck by the appearance of orderly and methodical progress. After Deryagin's visit to Nottingham in 1966 all this changed, and the activity shifted to various British laboratories involving people who were even at that time well known to me or whom I have gotten to know since. Therefore, for a while—until the main interest shifted once again, this time to the United States—I was right in the midst of the debate. At the same time, work continued in Moscow, and the output of research publications increased steadily.

Deryagin now also tried to interest British and American journals in his work, but such attempts proved unsuccessful. Two years later he was to complain about the reluctance of these journals to accept his manuscripts:

We have published much in Russian. We have had some difficulty in publishing this work, for instance in England. Two years ago I sent a remark to *Nature,* but the Editor refused it. In America we also had difficulties but there the difficulties had another character; some said it was not possible [credible?],[1] others said it had already been published by the Faraday Society, but I answered that this was nonsense because this was such a new phenomenon that to exclude all the preliminary parts would mean that nobody will understand it. This was the American *Journal of Chemical Physics.*

[1] Deryagin said "possible," but J. L. Finney, who edited the transcript, inserted "credible?". (Finney thought this was what Deryagin meant.)

From these remarks it is not immediately clear whether the refusal to publish was prompted by the astounding nature of the experiments or by the fact that the results were no longer considered original because the Faraday Society had already published them and nothing new had been added in the meantime. Be that as it may, it was several more years before Deryagin's work found its way into an American journal; by that time both he and his work had already received wide publicity in the West through other channels.

My own involvement with anomalous water began with a discussion in the spring of 1966 in Bradford, England. This was several months before the Nottingham meeting of the Faraday Society. The occasion was a symposium—of which I was organizer—on the physical chemistry of solvent mixtures in which one of the components was water. It was my first attempt at planning and organizing a scientific meeting, and I well remember being run off my feet most of the time, searching for spare bulbs for the slide projector, dealing with domestic problems and airplane reservations, and trying to keep the meeting running on time. It was a nightmare, and every evening I collapsed at home, grateful that no major derailment had occurred during the previous day. On one such evening I was brought face to face with the issue of anomalous water. My wife and I had invited several of the lecturers and participants to dinner, and afterwards over coffee Brian Pethica regaled us with an account of Deryagin's experiments.

Pethica was at that time director of the Unilever Research Laboratory at Port Sunlight, near Liverpool. The laboratory, staffed by some 800 scientists and assistants, was responsible for research and development activities in support of the company's worldwide soap and detergent interests. Prior to his appointment to the post of director Pethica had built up a most impressive team of physical chemists, devoted to the study of surfaces. He had the rare ability of injecting into a group of industrial scientists the intellectual curiosity and sense of urgency and professionalism so characteristic of many of the world-famous university research groups. It was, and still is, one of his favorite pastimes to propose apparently crazy hypotheses, shake accepted dogma, and ask deceptively simple questions that are very hard to answer. It was only natural that he and his colleagues should be intrigued by the puzzles set by Deryagin's experiments, especially since they lay squarely in their own domain of interest—surface chemistry.

Among the company assembled on that evening in my home the initial reaction was one of skepticism bordering on disbelief, and Brian clearly enjoyed the role of devil's advocate. The meeting broke up long after midnight, and I fell asleep reviewing all we had heard that evening. It was not surprising that I was plagued by repetitive dreams in which Deryagin's experiments were all mixed up with the more immediate problems of managing the symposium.

Also among the participants at the Bradford meeting was another scientist with an interest in Deryagin's modified water, although of a rather more personal nature. Alec Bangham was engaged in reading the scientific writings of his father, Donald Bangham, covering the period 1928–1948. It became clear to him that the phenomena described by Deryagin had already been discussed at some length by his father, who had reached the conclusion that marked differences existed between liquids in bulk and in the adsorbed state on solid surfaces. Over the years much of this work has been published in the *Proceedings of the Royal Society.* Alec Bangham was justifiably "affronted by Deryagin's failure to acknowledge my father's earlier work." He therefore felt motivated to draw attention to the fact that his father and others had extensively worked on and speculated about the phenomenon that was soon to be summarized by Deryagin in his famous paper at the Faraday Discussion of 1966 as "the ability of glass and quartz surfaces to change physical properties of many liquids to a great depth."

At the Bradford meeting Alec Bangham pointed out that the similarity between the properties of adsorbed water and the newly discovered anomalous water was too great a coincidence for them both to exist, and that if anomalous water was real it could presumably be prepared in very large quantities from low-grade coal, since millions of tons of water, adsorbed on coal, are ''mined'' annually.

Eventually, the brothers Alec and Derek Bangham published a letter in *Nature* drawing attention to the work of their father, and this brought to light a fundamental difference between the original work of Donald Bangham and the more recent experiments and claims of Deryagin: Deryagin maintained that the modified water, once formed, *was capable of existing independent of the solid surface,* whereas Bangham had ascribed his results to an adsorption of ordinary water at a solid surface. He had never believed that such adsorbed layers were formed irreversibly and could have an existence removed from the solid surface; that is, he discounted the structural-memory hypothesis. The realization of this perhaps subtle distinction eventually led Alec Bangham to send a letter of apology to Deryagin, a letter that elicited a very generous reply.

I have dwelt at some length on this seemingly minor personal aspect of the polywater controversy because apart from Brian Pethica it was Alec Bangham who first discussed with me the significance of Deryagin's work, but also because I want to illustrate that what scientists write (or do not write), and the manner in which they write, can have very personal repercussions. It is the convention in scholarly publication to make due acknowledgment to any theories or results originating from other sources, and perusal of the bibliography following an article sometimes throws an interesting light on the author's attitude and motives. I shall discuss the conventions and mystique of scientific publication in more detail later, but by way of explaining Alec Bangham's objections it is relevant to note—without comment at this stage—that Deryagin's paper in the 1966 Faraday Discussion proceedings contains a bibliography of thirty-six previously published works. Six refer to works by non-Russian authors, and of the remainder twenty-three refer to publications listing Deryagin as one of the authors.

The Nottingham meeting in the autumn of 1966 had provided Deryagin with new personal contacts. He was invited, nearly two years later, to report on progress at a small, select meeting of surface chemists at the Welsh resort Port Meirion. The Port Meirion meeting was less formal than the Faraday Discussion, and Deryagin was better able to explain in detail his experiments, their results, and his thoughts on anomalous water.

Several laboratories with active interest in anomalous water now extended invitations to Deryagin, in particular the Unilever Research Laboratory at Port Sunlight, the Cavendish Laboratory at Cambridge, and the Crystallography Department at Birkbeck College, London. This last contact was to prove particularly significant to the future of anomalous water. The head of the laboratory, John D. Bernal, had as a young man been very interested in the subject of ''water structure'' and in 1933 had been one of the authors of the publication that laid the foundations for most of the developments which have since taken place. Although Bernal's main activities had over the years turned to other fields, he had always maintained a keen interest in problems surrounding the physics of the liquid state.

Bernal was a remarkable man, an inspired teacher and research leader whose interests far transcended his immediate professional boundaries; he was an academic in the truest sense of the word. Like so many of his generation he had watched with great hope the aftermath of the Russian revolution, and the rise of Fascism and Nazism in Europe had led him to strengthen his connections with the Soviet scientists and institutions. This was made easier by his undoubted sympathy with the political and social policies then pursued by the Soviet Union, which opened many doors for him.

However, unlike some of his contemporaries, Bernal never allowed his political sympathies to cloud his critical faculties where science was concerned.

Deryagin was duly welcomed at Birkbeck College, and lectured to a very attentive and skeptical audience. In this audience was John Finney, a young Ph.D. student of Bernal's, who, on looking back, confesses that at the time he was not yet "fully aware of the nonideal nature of scientists and their work." It was to him that Bernal eventually gave the task of repeating Deryagin's experiments.

For obvious reasons, nothing from Birkbeck appeared in print for about two years. The practical problems posed by this type of research are formidable, as Deryagin had found. Any attempts at shortcuts—and there were to be many of these later on—inevitably laid the investigator open to charges of contamination by impurities. The years 1967 and 1968 were a time of germination, but all the while the Russians continued to be active, further elaborating the remarkable properties of anomalous water and putting previous experiments on a sounder footing. It was discovered by the Moscow group that anomalous water boiled in the neighborhood of 200°C, but the vapor was apparently stable at temperatures close to 450°C. This added a further puzzle to the growing list: How could it be that the vapor in contact with anomalous water maintained the strange properties of the liquid species? "Ordinary" liquid water may be quite atypical in its behavior, but at least in the vapor state it behaves much as one would expect it to behave from a knowledge of its simple chemical composition.

The progress in experimental studies of water is dotted with claims of anomalies, and Fedyakin's liquid columns in capillaries were just one more example of such apparent anomalies in the behavior of water under the influence of various extraneous factors. There is, for instance, an extensive literature dealing with the effects produced on water by magnetic fields. It has been claimed that magnetically treated water does not wet hard surfaces very well, and that this effect can be used in the processing of mineral ores by flotation. Then again, such magnetically treated water has been claimed to reduce the stability of colloid suspensions, an effect exploited in the preboiler treatment of industrial water. Most of these claims originate from the Soviet Union. Deryagin, in a critical assessment of these particular claims, wrote the following:

. . . in spite of the extensive industrial use of magnetic treatment the cause of the special properties thus acquired by water have not yet been investigated. The vagueness of the factors controlling this effect results in that it is not always reproduced the same. Therefore, the experimental data of different authors are often in disagreement, and in some cases they are actually contradictory.

The chief conclusion from analysis of the experimental data is that dormant water, unlike running water, does not demonstrate an appreciable change in the properties after remaining in a magnetic field . . ."

Similar critiques could probably be written about other claims of strange effects that have from time to time been made about the influence of external factors on the properties of water. In 1968 Paul Barnes, then at Birkbeck College, had recently completed his Ph.D. thesis at the University of Cambridge on the properties of ice surfaces and had come across what he thought were "anomalous viscous effects." Also at Birkbeck College, a visiting scientist from Pakistan was investigating claims about anomalous effects of temperature on the viscous flow of water and aqueous solutions. Anomalous water possibly provided a clue to such experimental observations. Bernal's long involvement with studies of liquids and his close contacts with Soviet science, coupled with the experiences of Paul Barnes at Cambridge, made it entirely reasonable for Barnes and Bernal's student Finney to set to work in an attempt to verify Deryagin's claims and possibly to explain the origin of the effect. They were helped by a detailed discussion with Deryagin, which was taperecorded. The following excerpts from the transcript illustrate some of the conceptual problems and paradoxes raised by the existence of anomalous water:

Bernal: What is your theory of the genesis of this type of water?

Deryagin: This is the most unclear point of the work. We have only two points clear, that this water may appear on the condensation of vapor but not as a result of direct contact of liquid water with quartz. Liquid water does not transform in some of our kind of water even after prolonged contact even at raised temperature; but in the form of vapor it can be modified.

Bernal: Does this mean that the molecules of this form of water exist in the vapor?

Deryagin: No. I suggest that this form of associated molecules does not exist in the ordinary water and in the vapor of the ordinary water. This association takes place as a result of condensation only. Perhaps this may be explained. I offer this hypothesis, that in the process of condensation there is evolved some extra energy, extra latent heat of condensation.

At this stage John Finney was asked to summarize:

Finney: Deryagin is suggesting that there is an extra latent heat contribution which is providing sufficient energy to overcome some kind of activation energy which is required before you can form this kind of water . . . .

The interview continues:

Finney: If you try to condense normal water too quickly, does [modified water] not form?

Deryagin: There are two stages. The first stage is the process of quick condensation when the vapor is supersaturated. After this we decrease the vapor pressure, and then we are waiting, as a rule one or two days, and then the nucleus of [modified] water increases in quantity very slowly.

Finney: Your initial nucleus is formed then on your first condensation?

Deryagin: Yes. Then it grows. There are two stages, one short and quick nucleation, and the other very slow state of condensation in subsaturated atmosphere.

Bernal: Have you any idea what the structure is of this new kind?

Deryagin: No, we do not know. But it may be a ring [of H_2O molecules] or a square or it may be tetrahedral.

Bernal: When did you first discover this phenomenon?

Deryagin: The first observation that led us to this new branch of water science has been made, not by myself, but by physicist Fedyakin six years ago. My papers have referred to this. He remarked that in one column of water with methyl alcohol or some other liquid inside a capillary, a glass capillary, not quartz, and taken for a prolonged time near the initial column, have formed two delta columns [the secondary condensation columns in which anomalous water was first obtained]. They grew during many weeks. This was a very slow procedure because this phenomenon was observed not in vacuum but in the presence of air and with diffusion of molecules of vapor through the air the process of formation of delta columns is very slow, but it is the great value of Fedyakin that he concluded that the water in the delta column must be in some different state than the ordinary water, because otherwise it is not possible to explain why the water distilled from the primary column to such delta columns. He realized that there must be some significance, but Fedyakin did not realize that the cause lies on a molecular level. His opinion was that it was the result of some different structure of the water only, like some crystals, but when we began to work with him and with my other pupils and colleagues, then we realized that the basis of this phenomenon lies on the molecular level and consists in the formation of such associated molecules. We have been working on this for some four to five years.

Bernal: In my opinion this is the most important physical-chemical discovery of this century.

Deryagin: I am very glad to hear you say this. I would like to ask you something. Would it be possible for you to write something later about your opinion on the significance of this work, as you are the principal specialist on the physics and chemistry of water? It would be very important for me to get such an estimate.

Bernal: I will be glad to do this. But I would like to ask another question. In your view, what is the biological importance?

Deryagin: We are beginning to study this problem, working with one of our pharmacologists. There are many very striking phenomena of the effect of very very small concentrations dissolved in water of some compounds, and such small concentration acts, for instance, on the heart of a frog; the heart of this animal is exceptionally sensitive to some compounds in extremely low concentration, one molecule per cubic millimeter or even more dilute concentration, and perhaps we shall subject the heart of this animal to investigate the biological effect of this water. Now there is another question that I have been accustomed to hear, that is, where may this water be present in nature? And to deal with this question I asked an astrophysicist in my country, in Estonia; he expressed the supposition that this kind of water is present in the "argent" clouds—silver clouds.

Bernal: Very high.

Deryagin: Yes, a height of 80 kilometers. Ordinary water would be crystallized. The reason why this astrophysicist put forward this hypothesis is that the light diffused in these clouds when analyzed has proved that these clouds consist of drops of size about one micron. Ordinary water cannot be in such a state because, first, there is high cosmic vacuum there and water would evaporate and, secondly, this water would crystallize. Perhaps this water we are discussing may be the constituent of these clouds. This supposition puts another question. There may be some other mechanism for the formation of such water, a mechanism that uses conditions that are there, cosmic and other kinds of radiation. It is possible there are conditions for such formation. There are no conditions for the formation of ordinary water.

Finney: Are you talking in terms of water which is partially modified and water which is fully modified, and so presumably most of the time you are dealing with a mixture of water?

Deryagin: Yes. We always prepare a mixture of water, partially modified, but after we continue the evaporation it would become strongly modified water. We can also do the inverse. We can take strongly modified water and mix, not very quickly, sometimes it lasts one week, because we measure the coefficient of the diffusion of the molecules of this modified water in ordinary water. It is about five or ten times lower than the coefficient of self-diffusion of ordinary water. So the diffusion is a slow process.

Finney: Can you actually say you can prepare water which is 100% modified or do you not know this? Can you say: This water is 100 percent modified? Have you any way of telling, any way of measuring the degree of modification?

Deryagin: Well, the degree of modification is not so precisely measured, but we have many indicators of the degree of modification. Firstly, the more convenient is the refractive index. Then, for instance, the vapor pressure: The more convenient is to investigate the dilatation at low temperature.

Bernal: You should find a name for this modified water.

Deryagin: We intend to wait a bit because in some months' time things will become clearer and we may then give a name to this water, polymerized or tetrameric. At the moment it is not quite clear what it does consist of. We will investigate other properties of this water, perhaps that will contain higher associated molecules. Then we must say something about the polymerization of the water. When it is not so, then there are only tetrameric molecules and then we can name it tetrameric water. We cannot at present make a choice between the different possibilities.

Bernal: I understand you very well. I agree that trouble should be taken to find a correct name for it.

Here we have all the ingredients of what soon was to become a *cause célèbre:* the ''most important physical-chemical dis-

covery of the century," anomalous water in the frog's heart
and in the upper atmosphere, and even the suggestion of the
name by which this substance was soon to become known.
Such speculations were quite in order in the privacy of
Bernal's office. It was quite another matter when, eighteen
months later, interviews based on hardly any better or more
comprehensive evidence than was available to Deryagin
(who, as a conservative man, was not given to making flam-
boyant claims) began to appear in the press. Whereas
Deryagin intended "to wait a bit because in some months
things will become clearer," others were not so disposed.

At the time of the Bernal-Deryagin discussions it was already
agreed by those active in research on anomalous water that
priority had to be given to devising ways of preparing the
substances in quantities larger than the trace amounts that
accumulated in the narrow capillaries, as in Deryagin's
method. It was also apparent that, despite the activity in
Moscow, this scaleup was still eluding the Russian scientists.

In addition to his two visits to Britain, Deryagin had in 1967
the opportunity of lecturing to an American audience, at the
Gordon Research Conference on the Chemistry of Interfaces,
held at Meriden, New Hampshire. The Gordon Research
Conferences bring together a limited number of front-rank
scientists in a given scientific area for discussion of recent advances and exchange of views in an informal atmosphere.
One of the conditions is that nothing that is said will be published or even be cited. The organizers of the Gordon Conferences usually aim to mix experienced scientists with some
younger ones, who are expected to benefit greatly from a
week of intense, high-level scientific debate. As a leader in
his field, Deryagin had on several occasions been invited to
these conferences, but had never actually been able to accept. One cannot but speculate whether this was due to personal inclination, refusal of permission, or the workings of
the Soviet bureaucracy so expertly described by Zhores
Medvedev in *The Medvedev Papers*. Whatever the reasons,
Deryagin did accept the invitation in 1967, and at Meriden

he was able to meet and lecture to a number of the world's leading surface scientists.

In character and atmosphere the Gordon Research Conferences are the very antithesis of the Faraday Discussions, where the accent is on discipline and decorum. The Gordon Conferences make a point of informality in both organization and dress. The lecture program is flexible and subject to change at the last moment. Scientific discussions take place during the morning and the evening, with the afternoon left free for swimming, sailing, tennis, or unstructured, spontaneous discussions. Traditionally the conferences take place at a number of private schools in the beautiful countryside of New Hampshire. These schools are modeled on the famous Public Schools of England, with minimum levels of personal comfort. Compared with other scientific meetings in the United States, which are often held in plush hotels, the Gordon Conferences are spartan in the extreme. Yet they enjoy unequaled prestige, and there are few scientists who would not jump at the opportunity to spend a week in the un-air-conditioned heat of a New Hampshire summer, sleeping in an excruciatingly uncomfortable bed and enjoying washing facilities that compare with those offered by the majority of European railway trains.

Since local regulations forbid the sale of alcohol at the schools used by the Gordon Conferences, every participant has to import his own supply, which is kept for him in the bar. It is in the bar that the scientific debates continue long after the official evening session has come to an end. The previous day's lectures are reviewed and discussed, and lecturers are subjected to cross-questioning by skeptics. New theories are born and old ones put to rest, and the seeds are sown for future collaboration between scientists who live and work thousands of miles apart.

The Gordon Conference rules state that only original, unpublished material must be presented, but this rule may be bent when the organizers feel that an important issue that has not previously received enough publicity is at hand. So it was with anomalous water: Deryagin was encouraged to re-tell his story, although it had already been widely published in the Russian journals. By all reports the response was mixed. The majority of those present did not want to know, and most of the remainder listened with amused tolerance. A handful were prepared to give anomalous water a second thought, but no more than that. As a means of convincing and inspiring American surface chemists, this particular Gordon Conference proved a failure. The convincing and inspiration were eventually achieved by a very different route.

It is debatable whether the lack of response was due to disbelief or to a lack of understanding of the implications of Deryagin's work. The lack of understanding had been obvious during the Faraday Discussion of the previous year. I well remember that during an extended visit to the United States in 1968 I was on several occasions asked about progress on anomalous water. These questions were always couched in the form of jokes, tinged with heavy irony. It was obvious that nobody was prepared to believe a word of it.

Brian Pethica told the participants in the 1968 Gordon Conference on the Chemistry of Surfaces of the successful repetition of Deryagin's experiments by his colleagues at Port Sunlight. Once again the general reaction was one of justifiable skepticism, but on this occasion at least one scientist returned home determined to take a close look at anomalous water. He was Robert Stromberg, of the U.S. National Bureau of Standards in Maryland. Was it that Brian Pethica was more persuasive than Boris Deryagin, or was it just the repetition and confirmation of last year's story that led Stromberg to propose to his superiors that he be permitted to investigate the new form of water? Whatever the case, this decision was one of the major developments in the polywater episode, and Stromberg was the first American to take anomalous water seriously enough to repeat and extend the work of Deryagin, Pethica, and Bernal.

In the meantime, news of the experiments at Birkbeck College and Port Sunlight had spread among surface chemists. The wider scientific public had been alerted to anomalous water in September 1968, when the letter by the Bangham brothers was published in *Nature.* This letter spelled out in more detail the problems involved in the study of liquid films near solid surfaces, and its title explicitly stated the fundamental principle: "Very long range structuring of liquids, including water, at solid surfaces." *Nature* has a very wide circulation, and it was to be only a matter of time before Western scientists would become aware of what was going on.

Probably the first reference to anomalous water in the mass media was a report in the scientific section of the *Frankfurter Allgemeine Zeitung* of May 1, 1968. The article, headlined "Water with density 1.3," summarized the Russian experiments and asked some very pertinent questions: Could there be anomalous water in other capillary systems, such as biological tissues? How was it that "Deryagin water" could remember its origin after a long period of time and after having been pushed along the capillary for a fair distance? The report was followed a week later by another with the news that the Deryagin group had now succeeded in producing anomalous water without the need for a narrow capillary, and that its molecular weight was found to be 72, which corresponded to the chemical formula H_8O_4. This was the first suggestion of a water polymer. In the second report the feature writer allowed himself some degree of speculation: It was now reasonable to assume that anomalous water did have an existence in nature. After all, porous, quartz-containing minerals were common. Perhaps "Deryagin water" was the reason why it was so difficult to completely dry many materials. Perhaps this kind of water also existed in the atmospheres of planets where water vapor could hardly be detected. Here was the first reference to polywater on other planets.

The British and American news media did not discover anomalous water until 1969, probably because of their parochialism. There were, however, isolated reports, most of them based on garbled thirdhand accounts of the Russian experiments. I recall a telephone conversation with an executive of a large adhesive company. Word had reached him of a kind of water that had a very high viscosity—just the sort of property one would look for in the formulation of adhesives. He had visions of completely new products based on this water and was extremely disappointed to learn that this wonder water only came in microgram quantities at best. There were other similar calls and letters from industry and hospitals, but on the whole there was still a general unawareness in the British and American scientific communities. In a way this was not too surprising. Scientists tend to be specialized and very selective in their reading and in their professional acquaintances, so there was little chance that many of them might have been exposed to what was at the time almost a minor aberration of surface chemistry. The written accounts, if they were available in English at all, were in the jargon of the surface chemist, and the meetings where anomalous water was discussed were attended almost exclusively by surface chemists.

How then did the subject of anomalous water become respectable in the United States? The man more responsible than anyone else was Ralph Burton, then a liaison scientist for the U.S. Office of Naval Research (ONR) in London, whose job it was to write regular summaries of research in Europe for distribution to scientific establishments in the United States. These summaries were based on what he saw and heard during visits to university departments and at scientific meetings, and on what he found by careful reading of the press.

The attitude of British scientists toward ONR liaison scientists was an ambivalent one. The U.S. Navy was providing generous funding in support of many scientific ventures in British universities, but the activities of such observers were regarded as hardly more legitimate than spying. It was well known that news of very recent results often found its way to the "competition" (American university departments working on similar areas) long before the scientists concerned felt ready to publish their findings. It was even rumored that there had been instances where American scientists had benefited from such advance information to the extent that, with their superior resources, they could complete a particular piece of work and publish their results before the British originator had a chance to do so. I have no reliable way of judging how justified such attitudes might have been, but they were quite widespread. As a result, some British university departments had an unwritten policy to keep out of sight anything of a sensitive nature when the "observer" from the ONR was visiting the laboratory.

By 1968 Ralph Burton must have felt that anomalous water was of sufficient interest to warrant its inclusion in his reports. He had been in contact with the scientists at the Unilever Laboratory at Port Sunlight and with Bernal and his colleagues. Also, the press department of the Soviet embassy in London had used the occasion of Deryagin's visit to Bernal's laboratory to issue a press release on the subject. John Finney recalls that on the strength of this activity Ralph Burton treated him to a "rather splendid lunch at the U.S. Embassy."

The first ONR report to Washington summarized Deryagin's earlier work and described it as "thought-provoking." At that time, however, it had not yet provoked much thought among those who should have been the first to be alerted. Burton also noted the skepticism that had greeted the Deryagin disclosures. In subsequent reports he summarized the properties of the fluid as they had been gradually appearing in a succession of Russian publications. By cultivating those British scientists who were actively involved in the study of anomalous water, Burton was only doing his job. He sensed the hostile reaction to such advances when he wrote in his third report that the British scientists had "momentarily drawn a gentle cloak of silence around their activities

. . . to protect their new findings until these are ready to go into print.'' How right he was. Since I was known not to be a competitor, I was treated as a confidant by various factions, and there was a very definite anxiety that eventually the U.S. scientists would be able to ''scoop'' anomalous water. The reasons for this anxiety were not that anomalous water might have some military applications, but purely the fear that someone else would preempt publication of the definitive experiments.

Those in charge at the ONR reacted to Burton's detailed and persistent reports by calling a working symposium in Washington, D.C. during February 1969 in order to discuss ''anomalies in the properties of liquid water.'' Invitations were limited to American scientists. (Let us remember that Americans generally, but American scientists in particular, still had vivid memories of how the Soviet Union had beaten them with Sputnik I. Now there were fears that the Russians had scored yet again, but America was not to be beaten so easily a second time.) Making the symposium exclusive must have been a counterproductive measure, because at the time there were only two or three American scientists with any firsthand experience with anomalous water. The remainder only knew by hearsay or from Deryagin's lecture at the 1967 Gordon Research Conference, and they had shown themselves to be unimpressed.

An exception to the lack of excitement was Frederick Fowkes, at that time director of research for the Sprague Electric Company, who not only had a long-standing interest in surface chemistry but knew Deryagin and had followed his work on anomalous water right from the beginning. Since he had attended the IUPAC meeting in Moscow and the Faraday Society Discussion in Nottingham, he was in a better position than most to gauge the response to Deryagin's work. For all these reasons Fowkes was given the task of summarizing the ''state of the art.'' To judge by the enthusiasm with which American scientists became converted to the investigation of anomalous water and the readiness of the various funding agencies to make available considerable sums of money, he must have done a splendid job. The Washington ONR symposium certainly laid the groundwork for most of the activity that followed over the next three years, in which American scientists played the leading role.

He who has not lost his head about some things has had no head to lose.

Jean Paul Richter

4

Polywater Becomes Respectable

In 1969 the West began to cut into the near monopoly of the Surface Chemistry Laboratory of the Soviet Academy of Sciences in Moscow. Almost from the start of Western involvement a set pattern developed in the approach of the scientific community to the subject of anomalous water. There were those who performed careful experimental or theoretical studies and reported them through the normal scientific media. Then there were others—often on the fringes of the main effort—who, without themselves spending too much time and thought, came up with bright ideas on anomalous water's structure. The climate at the time was such that editors of various journals were prepared to accept the kind of hypothesizing and speculation that should never have survived the reviewing stage. Then there were the popular science publications, which carried digests of the articles in the more serious journals together with press interviews given by some of the more extroverted personalities who considered themselves the experts at the time. Finally, there were the mass media—daily newspapers and weekly and monthly magazines—which became aware of anomalous water in the early autumn of 1969. Stories could be found as news items or in the sections devoted to scientific and cultural affairs. In several instances even the editorialists got into the act. Anomalous water made the editorial page of the *New York Times* on September 22, 1969, right alongside the more familiar themes of crime and welfare.

In 1969 the issues were still quite clear-cut: The first and foremost order of business was to confirm the experiments of Deryagin and then to enquire into the nature and origin of anomalous water. The first serious attempt at the former appeared in *Nature* of April 12. Brian Pethica and his colleagues at Port Sunlight were now ready to summarize the past eighteen months' work. There was little of a revolutionary nature. The appearance and behavior of anomalous water were confirmed. Figure 5 (taken from the *Nature* paper) is a photomicrograph of the liquid columns that formed in the capillary tubes after they had been treated according to the Deryagin method. Figure 6 shows the effect of temperature on the length of the columns; this is a measure of the thermal expansion coefficient. Whereas ordinary water exhibits the well-known minimum at $4\,^\circ$C, the anomalous columns contracted in length right down to about $-8\,^\circ$C, and there was no evidence of freezing. However, the Unilever scientists were unable to find species with molecular weights higher than that of H_2O, which is to say 18. Also, attempts at a spectroscopic characterization of the liquid in the capillaries failed to reveal the presence of anything other than ordinary water. The authors concluded that the liquid was probably ordinary water with a small admixture of anomalous water, and the last sentence of their report in *Nature* reads: "The existence of 'anomalous' water as a stable and distinct molecular entity will only be established when it becomes available in amounts sufficient for an unequivocal characterization." This is a statement of the obvious, yet during the whole anomalous-water episode *nobody* ever prepared the liquid in sufficiently large amounts. Nevertheless, many scientists not only became convinced that it did exist as a "stable and distinct molecular entity" but were ready to pronounce on its molecular structure.

What I have called the "first serious attempt" by Pethica's colleagues had been preceded by some of the scientific speculators, who took Deryagin's results for granted and advanced some more or less plausible structures on the basis of very tenuous arguments. The editors of *Nature* and *Physical Review Letters* saw fit to publish such metaphysics, a judgment that even then was questionable and caused raised eyebrows among some of the experimentalists. (Perhaps the motivation was to stimulate scientific debate.) It is hard to be-

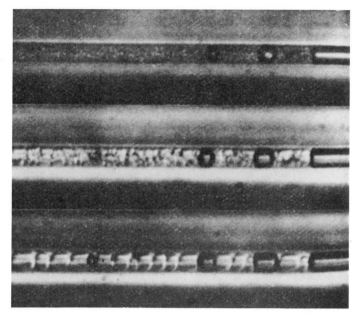

Figure 5
Photomicrographs of anomalous-water columns grown in 10-μm-radius Pyrex capillaries, at −27°C (top), −14°C (middle), and −5° (bottom). Reproduced, with permission, from Willis *et al., Nature* 222 (1969):159.

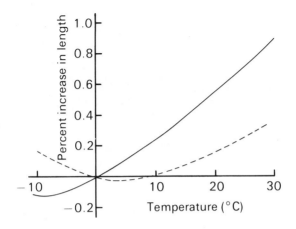

Figure 6
Thermal expansion of capillary columns of water and anomalous water. Apart from the difference in the temperatures at which the minima appear, note also the steepness of the anomalous-water curve; it indicates a coefficient of expansion $1\frac{1}{2}$ times that of water.

lieve that the work described in *Chemical Physics Letters* by Stig Erlander could have taken much more time and effort than the writing of the article. One of these early speculative articles, by Richard Bolander, James Kassner, and Joseph Zung from the University of Missouri, contained the first acknowledgment of financial support by the U.S. Office of Naval Research.

Also in the spring of 1969 there appeared the first of a long line of popular science articles, most of which had catchy headlines. This article, "Polymerized Water—Is It Or Isn't It?," signed by none other than John Finney of Birkbeck College, illustrates the ambivalence of serious scientists toward the strange fluid. Many of us wanted to believe that it existed, but we did not really dare to do so. Even if we made the leap of faith, we did not dare to testify in print. Writes John Finney: "By not fitting in with existing ideas, anomalous water could trigger off an important advance. . . . If only we could make a thimbleful, the problem would very quickly be resolved." The following three years were to see many variations of these two themes.

Chemical & Engineering News was quick to spot the article by the Unilever scientists, and reported an interview with Pethica which did not materially add anything to what had already been printed. Their reporter concluded that "Dr. Pethica and other scientists working in the field are keeping quiet about their experiments with bulk production of anomalous water." Quiet indeed, since there was little to talk about.

The central problem in the identification of a new chemical compound is the determination of its molecular structure, the linkage structure of the various atoms in the molecule, the length and strength of the bonds between the atoms, and the general shape of the molecule. This type of information is basic to an understanding of how such a molecule can interact with other molecules, how it takes part in chemical reactions, how the substance crystallizes on freezing, and how the molecules might interact in the liquid state. Answers to such questions can be gained through the study of how the molecule absorbs electromagnetic radiation. Such absorption can take place in different parts of the electromagnetic spectrum—from the short-wave, high-energy x rays through the ultraviolet and visible-light region to the long-wave infrared and microwave region. Collectively, such analytical methods are termed spectroscopy, and, when used in certain combinations, they provide powerful tools for the elucidation of molecular structure.

One of the most commonly used spectroscopic techniques relies on the observation that chemical bonds between atoms vibrate, much like spiral springs, and that the energies and amplitudes of such vibrations are unique to particular bonds in particular molecules. When a substance is subjected to infrared radiation, it will absorb such radiation at certain wavelengths corresponding to the particular energies of vibration of the various chemical bonds. The remaining part of the beam is not absorbed and will therefore pass through the substance unchanged; its intensity can be measured and a wavelength spectrum constructed. This spectrum will exhibit peaks corresponding to the wavelengths at which the molecule has absorbed radiation. This is shown for the water molecule in figure 7. Actually, the measurement of infrared absorption is complicated by certain experimental problems; the same (or very similar) information can often be obtained by Raman spectroscopy. (Figure 7 is in fact the Raman spectrum, but the peaks correspond to the absorption of energy by the chemical bonds—in this case, the O—H bonds—of the water molecule.)

Infrared and Raman spectroscopy are two of the most frequently used methods for identifying an unknown chemical material. Complete spectra can be stored in computer files, which can be searched for a spectrum identical or similar to that of the unknown compound, much as fingerprinting can be used for identification purposes. I shall return to the de-

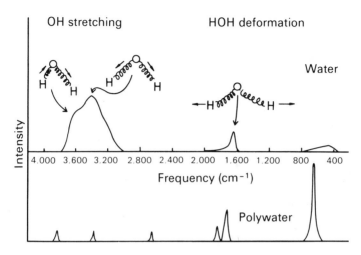

Figure 7
Raman spectra of water and polywater. The two O—H-bond-stretching vibrations and the HOH-angle-deformation band, characteristic of the H_2O molecule, are shown; they do not appear in the polywater spectrum. The frequency is the reciprocal of the wavelength of the radiation.

tails of the water and polywater spectra, because they were
the basis on which the proposed molecular structure of polywater came to be based.

The first spectroscopic study of anomalous water was published on May 24, 1969. The authors included a British
team from the Explosive Research and Development Establishment of the Ministry of Technology and an American
group from the University of Maryland. The man responsible
for this collaboration was Lionel J. Bellamy of the British
Ministry of Technology, who had succeeded in producing
enough anomalous water for such a spectroscopic study and
had enlisted the American spectroscopist Ellis R. Lippincott.
Lippincott had been professor of chemistry at the University
of Maryland since 1955 and director of a newly formed Center for Materials Research since 1967. He was a consultant
to the National Bureau of Standards, the National Institutes
of Health, and the Office of the Secretary for Defense. The
list of his honors was impressive, including awards from the
U.S. Department of Commerce, from the Gulf Research Corporation, and from the American Chemical Society, which
awarded him the Hildebrand medal in 1964. This turned out
to be particularly ironic because Joel Hildebrand, one of
America's premier chemists, was to take a quite uncompromising stand against all his fellow scientists who had been
naive enough to take Deryagin seriously. If any one name
was to become coupled with that of Deryagin, it was that of
Lippincott.

The Anglo-American experiments showed that the spectra of
anomalous water were quite different from those of ordinary
water, and the workers were convinced that they were dealing with a chemical that was "a new form of water and not
the result of casual contamination." As to the nature of this
new chemical, Bellamy, Lippincott, and their colleagues proposed that anomalous water "*must* consist of polymer
units. . . ." It was the first time that such claims were
backed up by convincing experimental results. About the actual structure of such polymers they could only speculate,

but they came down in favor of squares of four water molecules resulting from the influence of the quartz surface on the neighboring H_2O molecules. Such a hypothesis immediately raised new questions, because the structure put forward was completely at odds with the known structure of ordinary liquid water, which is based on the tetrahedron shown in figure 2a. However, Bellamy and Lippincott were not too worried about this, and stated that "the requirement for nominal 90° angles for water [instead of 109°] in this model does not seem to us to be a major difficulty in view of the very wide range of angles which commonly occur [for water] in crystals."

These new developments gave us all something to talk about during the summer of 1969, and it is only natural that several informal conferences were convened with the purposes of exchanging information and discussing what to do next. One such meeting took place at the University of Bristol, at the initiative of Douglas H. Everett, professor of physical chemistry. His department was and still is one of the leading centers for the study of surface chemistry, and Bristol therefore seemed a natural place for such a gathering. It was a very informal affair; the invitation came by telephone, and as far as I am aware, no written record of the meeting was ever prepared. The participants were asked to describe their own experiments, and this was followed by a critique of the available Russian reports. I cannot remember that anything startling came to light, except that Bellamy hinted that at his laboratory a method had been developed for producing anomalous water in larger quantities. When pressed, he was most reluctant to reveal details, and it was clear that the presence of the ONR observer at the meeting was inhibiting free and open discussion. Perhaps it was this experience that prompted Burton's remarks about "gentle cloaks of silence" being drawn around the activities of British scientists. It was indeed a pity that the informal meeting at Bristol happened to take place just a few weeks before the publication of some new results which proved to be the most important single development since Fedyakin's 1962 discovery of liquid columns in capillary tubes.

Under the title "Polywater" Lippincott and Stromberg, together with two junior colleagues, published a substantial account in *Science,* the organ of the American Association for the Advancement of Science, whose emphasis is very similar to that of *Nature;* it publishes short, up-to-date articles covering the whole spectrum of science. The new revelations of Lippincott and his colleagues in the U.S. Bureau of Standards went far beyond the preliminary results Lippincott and Bellamy had published only a few weeks earlier. Now it was stated quite explicitly that the properties of the liquid in question "are no longer anomalous but rather, those of a newly found substance—polymeric water or polywater." Since this was the publication that acted as catalyst for most of the activity that followed over the next two years, I will quote some of the conclusions that were based on spectroscopic examinations of the by-now-familiar liquid, performed by one of the world's most renowned spectroscopists.

It is now necessary to discuss in more detail the nature of the infrared and Raman spectra of water. Essentially, three peaks can be observed, and they had been assigned to three modes of vibration of the molecule, as shown in figure 7. Two of these correspond to the stretching of the O—H bonds, and one to the bending of the angle between these two bonds. (Actually, the spectrum of water in the liquid state is very much more complex, because of new spectral features which arise from the interactions *between* water molecules and which are superimposed on the feature which reflect interactions *within* the molecule.)

A most surprising result of the new spectroscopic investigations of anomalous water was the complete absence of any spectral feature corresponding to the normal stretching vibration of the O—H bonds in ordinary water. Instead, several new bands appeared which seemed to have no place in the known spectrum of water. Figure 7 includes the spectrum of anomalous water, and the divergence from the water spectrum is indeed striking.

On the question of impurities, Lippincott and co-workers stated that there was no evidence in the spectra for any contamination by silicon-containing compounds (which might have originated from the quartz capillary tubes) or by oils, greases, and the like (which might have found their way into the samples from the vacuum systems used).

A computer search of 100,000 spectra convinced Lippincott and his colleagues that their spectrum was in fact unique and did not resemble the spectrum of any known substance. Previously proposed structures were therefore considered ''less than adequate'' to explain the physical behavior of the liquid, and the structure now proposed was based on a ''previously unrecognized type of bonding for a system containing only hydrogen and oxygen atoms.'' (I shall return to this type of bonding and its structural implications in the next chapter.) The structure favored by Lippincott, which apparently also accounted for all the physical properties painstakingly measured by Deryagin's colleagues, was one based on a polymerized water—water molecules forming long chains or hexagonal rings, or a mixture of both (see figure 8, p. 93).

There is evidence here of schizoid behavior on the part of Lippincott, who just a few weeks earlier had gone on record as favoring the idea of squares of water molecules. Surely the spectroscopic studies on which the two publications were based must have been performed at much the same time, certainly in the same laboratory, although on different samples of anomalous water. Yet Lippincott now referred to the recent paper in *Nature* as though it had been written by completely different people.

The climax of the arguments of Lippincott and colleagues— the punch line, so to speak—was the following: ''We believe that water is restructured on the . . . quartz into a polymeric form with bonds and bond energies completely different from that of ordinary water. Its properties are not those of water, and it should not be considered to be or even called water. . . ." This was strong stuff indeed. Little wonder that it set the tone for what was to follow. Nevertheless, whatever criticism can be leveled against the quality of Lippincott's work or the manner of its presentation, at least the reader was left in no doubt as to the stand taken by the authors. There are no question marks, no ambiguities, and no equivocation. This is one of the rare examples in the chronicle of polywater where a group of authors took a definite stand based on their reading of the experimental evidence.

Before leaving the climactic experiments of Lippincott and Stromberg, we should note that reference is made in their text to a very sensitive type of chemical analysis which did in fact detect minute amounts of foreign materials, both in the quartz capillaries and also in the liquid condensed in the capillaries. However, the amounts of silicon and sodium found were so small that they could hardly have influenced the formation or behavior of the polywater. They could certainly not account for the appearance of the spectrum that differed so radically from that of ordinary water. The question of sodium impurities in quartz and their possible role in the formation of anomalous water had been raised by Frederick Fowkes, and was to receive more attention by Arthur Cherkin later in the year. These and other related activities might not have been known to Lippincott at the time of his own experiments, although the question of quartz contamination had certainly been raised at the ONR symposium in Washington, which Lippincott had attended.

No one who read the Lippincott polywater paper in *Science* of June 27, 1969 should have been surprised that its controversial nature sparked furious activity and vociferous debate in the American scientific community. The popular science journals and the newspapers fell over each other to report on polywater, and much inventiveness was shown by caption writers. In the wake of the original work there arose a veritable army of "experts" whose main functions were reinterpre-

tation, commentary, and the writing of learned and sometimes erudite reviews in various specialist publications. Some
of the authors remained anonymous, and others might later
have wished to have remained so. Such reviews usually consisted of a rehash of Deryagin's findings supplemented by
Lippincott's hypotheses, and generally ended up with speculations on potential applications of polywater, perhaps in the
development of a steam automobile, as a moderating fluid in
nuclear reactors (heavy polywater), or as a lubricant or a cor-
rosion inhibitor. Then there were never-ending suggestions
on the role played by polywater in life processes, a subject
that provided much nourishment for fertile imaginations.

Yet others in the Western press were dissatisfied that an important phenomenon had ostensibly been discovered in the
Soviet Union. That could be corrected in one of two ways.
The direct method of rewriting history is illustrated by a
headline in the *Miami Herald* of July 30, 1969: "Miami
Scientific Team Creates Mysterious New Form of Water."
The article below this headline made no mention of Lippincott, let alone Deryagin, although careful reading of the story
does reveal that polywater "was accidentally discovered,
probably [sic] by a group of Russian scientists." The more
subtle way of searching for a non-Russian originator of polywater is exemplified by an "in depth" account of polywater
in the *Saturday Review* of September 6, 1969. Here we are
told about an American chemist, Walter Patrick, who, inci-
dentally, had invented the process on which the functioning
of the gas mask is based. Patrick had apparently discovered
that when silica gel containing water was subjected to heat,
some of the water was retained in the solid even at temperatures at which it should have boiled away. One of Patrick's
students at Howard University, a Russian immigrant named
J. L. Shereshefsky, took up the general problem posed by
liquids confined in narrow capillaries for his Ph.D. thesis,
which when published in 1928 did not create much of a stir.
Shereshefsky found that water in capillaries has an even
lower vapor pressure than that predicted from its curvature

by the Kelvin equation. Next the story shifts to a Soviet physicist, K. M. Chmutov, who in 1937 is reputed to have repeated and verified Shereshefsky's experiments. From Chmutov to Fedyakin in 1962 is then only one step. Where does Bangham fit into this scheme? This manner of tracing developments back to their true origin is fairly pointless, because discoveries are not planned and cannot be legislated. Also, their real sources are often hard to track down. Quite apart from this, Fedyakin, working on his own in faraway Kostroma, must have been completely unaware of the existence, let alone of the work, of one Walter Patrick. It is disingenuous to claim, even if only by implication, that Fedyakin's work constitutes a logical followup of Patrick's original ideas. (An ironic aside to this story is that Warren H. Grant, one of the four authors of the famous polywater paper in *Science,* did his graduate work at Howard University under none other than Shereshefsky, without ever finding out about his professor's former activities.)

Along with Fedyakin's initial discovery of anomalous water, which caused little stir, the report by Lippincott and his colleagues proved to be one of the two major events of the polywater period. They ensured themselves the widest publicity by lecturing throughout the United States and elsewhere—as far afield as Dubrovnik, Yugoslavia. Polywater rated a half-day session—an honor indeed—at a Gordon Conference on biopolymers. Deryagin reappeared in Florence and Sydney, and others retailed the story second-hand at conferences in Bombay and San Francisco. Further publicity was provided by the popular science journals. *New Scientist* devoted a whole page to a story entitled "Enter polywater—Amid Alarums and Excursions," which ended with the following words: "Ellis Lippincott is well known for the quality of his spectroscopic equipment, and for the quality of the results he produces from it. His apparently unbelievable suggestion that 'water'—or perhaps it would be better to say oxygen and hydrogen atoms—can exist in this extraordinary form, has to be taken very seriously.'' And so it was. *Scientific American,* after once again summarizing the evidence, asked: ''If polywater is so stable, why has it never been found in nature? Because of its greater density, does it exist at the bottom of the ocean? Is this material found in the body, and if so, what is its role in life processes? How does it affect the choice of water as the standard liquid on which so many scientific measurements are based? What potential uses would there be for a stable polymer of water if it could be made in sufficiently large quantities?'' On the subject of industrial application, *Chemical & Engineering News* had this to say: ''Discussion of practical applications today is considered bad form. Noted scholars have been cut dead by their colleagues for less.'' (Times have changed. Nowadays the opposite is almost true: Potential applications are supposed to generate discoveries.)

On September 11, 1969, at a meeting of the American Chemical Society in New York, Ellis Lippincott told a large audience yet again of his remarkable results and conclusions. This time the news finally found its way into the mass media. The following morning the *New York Times* and the Cleveland *Plain Dealer* carried stories on polywater. These were followed by similar accounts in European papers, such as London's *Daily Telegraph* and Paris' *Le Monde.* At exactly this time also appeared the long article by John Lear in the *Saturday Review,* which in turn caused polywater to be elevated to the editorial page of the *New York Times* as an example of unplanned Soviet-American scientific collaboration. Once again the apocryphal story of Walter Patrick of Johns Hopkins University and J. L. Shereshefsky (misspelled) was told, and the inference was drawn that the Soviet work developed naturally from these earlier American discoveries. The *Times* went on to say that now American and British scientists had elucidated the ''true'' nature of polywater. Finally this: ''How much earlier might these discoveries have been made if political tensions had not hindered scientific collaboration for so many years?'' To those of us standing on the sidelines it was hardly a matter of collaboration, but looked much more like a scramble to be first to "get into" polywater, solve its structure, and produce it in reasonable quantities—not necessarily in that order.

Newsweek on September 29, 1969 supplied yet more background information and speculation on the possible uses of polywater. By now only "a few skeptics believe polywater is actually a mixture of real water and the material of the capillary tubes, but many scientists are prepared to accept it as a truly new form of water." It was also revealed that the U.S. Office of Saline Water had become involved in the affair in a minor way, by investing $50,000 in the search for methods of producing large quantities of polywater. In fact, on September 11 the Office of Saline Water had issued a news release to the effect that large quantities of anomalous water had been produced. One must wonder what became of them. Incidentally, a similar claim was made in the *Naval Research Review* later in the year.

At about the same time several groups of scientists were working toward an understanding of the nature and functions of water in biological cells. This is a problem that even now is by no means solved to everyone's satisfaction; the battle lines are well drawn between those who see cell water as a kind of polywater (that is, distinct in its physical properties from the water outside the cell) and those who maintain that an equilibrium exists between the inside and the outside of the cell. This problem goes right to the origin of the mechanisms that govern the movements of salts and nutrients into and out of the cell, and is therefore of some fundamental importance. To be fair to all concerned, I do not recall any scientist working in this area ever claiming that cell water can be equated with polywater, but the climate in the autumn of 1969 was such that strangely garbled stories appeared linking these two subjects. Wrote *Newsweek:* ". . . at least one research group is studying living cells for evidence of the strange new compound. If they find it there, another interesting question will arise: what is the role of this unsuspected chemical in our life processes?'' Repeated denials by a member of this research group did not find their way into *Newsweek* or any other publication.

As the learned debate among the academics increased in vigor and volume, and more and more letters arrived on the desks of the editors of *Nature* and *Science,* there exploded the second bombshell—this time in the pages of *Nature.* F. J. Donahoe of Wilkes College in Pennsylvania uttered a warning to all those hard at work trying to produce polywater. He wrote—and *Nature* printed—the following:

I need not spell out in detail the consequences if the polymer phase can grow at the expense of normal water under any conditions found in the environment. Polywater may or may not be the secret of Venus's missing water. The polymerization of Earth's water would turn her into a reasonable facsimile of Venus.

After being convinced of the existence of polywater, I am not easily persuaded that it is not dangerous. The consequences of being wrong about this matter are so serious that only positive evidence that there is no danger would be acceptable. Only the existence of natural (ambient) mechanisms which depolymerize the material would prove its safety. Until such mechanisms are known to exist, I regard this polymer as the *most dangerous material on earth.*

Every effort must be made to establish the absolute safety of the material before it is commercially produced. Once the polymer nuclei become dispersed in the soil it will be too late to do anything. Even as I write there are undoubtedly scores of groups preparing polywater.

Scientists everywhere must be alerted to the need for extreme caution in the disposal of polywater. Treat it as the most deadly virus until its safety is established.

I have quoted at length from Donahoe's letter because it illustrates several facets of a scientist's mental makeup. One must assume that Donahoe intended his letter to be taken seriously. His reasoning was based on the fact that when several physical modifications exist of a chemical compound, in this case H_2O, then the one with the lowest vapor pressure is intrinsically the most stable one, and all others can be converted to this stable form under a given set of conditions. That is not to say, however, that such a conversion is likely to happen spontaneously. Many substances whose existence we take very much for granted are in fact metastable and could only be converted into the stable form with the aid of high temperatures, pressures, or catalysts. For instance, at room temperature a mixture of hydrogen and oxygen gases is unstable with respect to liquid water. However, the probability of such a gas mixture reacting spontaneously to form water is extremely low. The situation is quite different when the gases are mixed in the presence of some powdered metals: They react explosively to form the stable form, liquid water.

Donahoe argued, therefore, that polywater had a lower vapor pressure than ordinary water, and its formation might therefore inadvertently be catalyzed. In time, all ordinary water would then be converted to polywater. We were back with Kurt Vonnegut's ice-nine, which in Cat's Cradle puts an end to life on earth. Science fiction had become science possibility. Donahoe's letter illustrates that the climate in 1969, just after Lippincott's revelations, was such that opinion immediately became strongly polarized—there were the "believers" and the "nonbelievers." Donahoe had clearly come down on the side of the believers, and he took the existence of polywater to its logical conclusion.

Of course, the replies were not long in coming, again in Nature. (They were actually "scooped" by the Huntsville (Alabama) Times on October 19, 1969, a week before they appeared in Nature.) The rebuttals were signed by two key members of the British scientific establishment, John Bernal and Douglas Everett, and by their colleagues at the Universities of London and Bristol. Donahoe was taken severely to task for his "unduly alarmist and misleading" letter. In the first place, it was pointed out, ordinary water had existed on this planet for billions of years, and in close contact with quartz, under ideal conditions for the formation of polywater.

The authors once again emphasized the extreme reluctance of water to yield polywater (of which only a few millionths of a gram had so far been produced, and not for want of trying on the part of many experienced scientists), and stated that if polywater could indeed grow at the expense of ordinary water, then we should already be a dead planet. Bernal concluded with this admonition: "By all means draw the attention of scientists to the dangers of their work, but make sure it is a real danger before alarming everybody else."

I do not believe that Donahoe (in contrast with some of his contemporaries) was seeking publicity; he was no doubt completely sincere, if misguided. At the time, environmentalist hysteria was not yet a facet of everyday life. Bernal's words of censure to the professional prophets of doom and gloom are much more appropriate today than they were when they were written.

If Lippincott's claim that anomalous water was not water but some new polymerized fluid had excited the scientific community, Donahoe's warning of the dangers that might lie in store for us had a much wider impact. Once again the debate moved into the mass media, and this made objective and dispassionate discussion from then on almost impossible. The U.S. Department of Defense became involved by supporting attempts by Barry Brummer of Tyco Laboratories, Inc., of Waltham, Massachusetts, to produce large quantities of polywater. Brummer's picture duly appeared in the pages of *Der Spiegel,* where the story was given generous coverage. *Time,* not to be outdone, carried Donahoe's warning of the "threat to life," but not the rebuttal by Bernal. Lippincott, when asked for his opinion, thought that the danger was slight but agreed that polywater should be "handled with care" until scientists knew more about it.

Increasingly the weekly news magazines referred to the connection between polywater and water in biological tissues, although to anyone familiar with current developments these two problems were completely unconnected. For semiliterate science reporting, though, it is hard to beat a piece in the *Guardian*. I can do no better than to quote:

Russian experiments, now confirmed in the United States and in England, have uncovered a new form of water, denser and more viscuous [sic] than the product of the tap and with a boiling point about 500°C. Polywater was first made by condensing steam in narrow tubes of glass or quartz, but there may be easier methods, for a number of American teams claim to have found it in biological material. It seems that most of the water in brain and muscle resembles the stuff discovered in Russia. In particular, the water in these tissues has a regular and symmetrical structure, such as might be expected from a crystalline solid but not from a liquid. Not to put too fine a point on it, the water in our bodies seems to behave like ice—a mystery which will be kept for dissection another day.

Meanwhile an American scholar . . . suggests that polywater, if once let out of the laboratory will go on a wild rampage across the globe, transforming the cool clear liquid that we use for drinking and washing into polywater, thereby destroying all earthly life—and giving the scientists a bad name too.

This concoction of truth, half-truth, and untruth, illustrated with a cartoon and served up in a witty and erudite style by a "quality" newspaper, made it hard for the uninitiated reader to understand the principles at issue. Such, however, was the contribution of the media to the polywater debate. (There were notable exceptions; the *Washington Post* presented a well-balanced account of what was going on.)

In the meantime the activity had spread to Australia, where at a large congress of IUPAC Deryagin once more figured as guest of honor. For the first time some of his experimentally determined properties of polywater were beginning to be challenged. Over the next two years Australian scientists took a major part in the action, and here, as elsewhere, opinions were strongly polarized. Meanwhile, the theoreticians moved into polywater, suggesting new and to them more plausible molecular structures. Here again, reading the small print reveals the support of various U.S. government agencies for this work. The rationale for such an involvement was pertinently expressed by Keiji Morokuma of the University of Rochester: "Though the existence of the anomalous water is still questioned and will take time to be established and the models proposed are not concrete, it would be fruitful to begin to examine from a theoretical point of view whether such models could be of any possibility." This type of reasoning is not too uncommon in scientific circles. To the lay person it may sound crazy: Why bother with theoretical speculations, if the substance in question may not even exist? Should not its existence be established first? Only if it does have a real existence should one address the questions of how and why. In the next chapter I shall deal with the different philosophies that motivate experimentalists and theoreticians. In the meantime, suffice it to say that after spending large sums of money on computer time we are not very much nearer to understanding how even *two* water molecules interact when they collide; nor do we know the mechanism by which many water molecules interact to form "ordinary" liquid water, let alone polywater.

The year 1969 ended with the first explicit suggestion that polywater was probably nothing more than a solution in water of soluble quartz components from "anomalous" regions in the solid quartz. Arthur Cherkin, of the Veterans Administration Hospital in Sepulveda, California, put forward this suggestion in the last issue of *Nature* in 1969, after having had a previous manuscript rejected. Here again, uncertainty is revealed by the question mark at the end of the title: " 'Anomalous' Water: A Silica Dispersion?" Question marks were a common feature in polywater publications. Nevertheless, Cherkin was no idle speculator; he had extensive experience in studying the solubility of glass used for bottling intravenous solutions, and this immediately suggested to him a more plausible explanation of Deryagin's results: that under the right experimental conditions water would leach out soluble material from glass or quartz. In a letter to Alec

Bangham he wrote: ''in 1969 polywater become the golly-water of our lay press; may it become the follywater of 1970?''

Cherkin was too optimistic in his estimate—in 1970 polywater reached the zenith of its short history. The American Chemical Society staged a meeting devoted exclusively to this subject, a meeting that brought together the supporters of Deryagin and Lippincott and the rapidly growing band of critics. Of course, much of the work reported at the ACS Anomalous Water Symposium in the summer of 1970 had already been in progress during the period covered by this chapter, but in telling the story of 1969 I have tried to confine myself to events as they happened during that year. Events of 1970, as my subsequent correspondence with many of the scientists who took an active part confirmed, indicate that during 1968—1969 a surprisingly large number of active scientists dropped whatever they had been doing as though it was of minor importance to devote their thoughts and activities to the elusive polywater, along the routes charted by Deryagin and Lippincott.

It was almost like a little gold rush. If Deryagin was right, and if polywater could be encouraged to grow in large quantities, then there were worthwhile rewards for whoever got there first. If on the other hand polywater proved to be a mirage, then kudos would go to whoever struck the death blow by providing incontrovertible evidence. By and large, scientists are ordinary human beings who have been trained to apply objective analysis to the solution of problems. However, such training—administered by other, equally ordinary humans—does not eradicate much deeper-seated forces that motivate us to adopt particular courses of action under given circumstances. In the 1960s most scientists enjoyed considerable freedom in the choice of problems and work areas, and were much less tied down than they are today by managerial discipline and considerations of cost-effectiveness. It was not altogether surprising, then, that many of those with first-hand knowledge of the behavior of water and solid surfaces, and even some who could hardly claim any such experience but were willing to learn, should become convinced that poly-water presented a challenge that could not be turned down.

Bourgeois scientists make sure that their theories are not dangerous to God or to capital.

Georgi V. Plekhanoff

5

Experiment and Theory

It is now easy to discern the pattern that by 1970 had been set for further progress in polywater research, although at the time things looked quite confused. There were several main lines of approach taken by different groups of investigators. For example, the Russian school, under the firm and undisputed leadership of Deryagin, continued to probe the bulk properties of the mysterious fluid. Looking back over the published work from the Moscow laboratories one must be impressed by the apparently orderly progress. It seems to have been a well-planned effort. On the whole, the Russians were not much concerned with the mechanism by which polywater was formed, or with its molecular structure, or indeed with the establishment of a theoretical basis which could explain, in terms of what was known about chemical bonds, the remarkable stability and the physical properties of polywater.

By this time those in authority in Moscow had apparently realized the value of publicity. Deryagin's earlier work had been published in the Russian literature, of which only a small part was at that time translated into English. Now there was a spate of English articles, and Deryagin was being received as an honored guest at the international congresses. It is intriguing that, despite the military backing for research into polywater, no attempt was ever made by either side to stifle publication or discussion.

By 1970, few surface chemists had not heard of Deryagin and his modified water, but a careful reading of the experimental evidence was bound to raise certain doubts in some quarters. The individual results as published by the Moscow group did not fit at all well within commonly accepted theoretical frameworks. For instance, Douglas Everett and his colleagues at Bristol recalled that according to Deryagin modified water was usually formed as a solution in ordinary water, and that the ordinary water could then be removed by distillation, leaving a semisolid residue of modified water. In other words, polywater was soluble in ordinary water. But once this fact was accepted, then many of the seemingly remarkable attributes of polywater could be readily explained in terms of the properties of solutions. Similarly, if polywater did in fact contain silica crystals of microscopic dimensions, then again there would be no need to postulate the existence of a new polymer of water. Everett counseled that, before advancing new, somewhat revolutionary structures of polywater, it would be more sensible to try and account for the various documented results in terms of accepted (and acceptable) theories. The philosophy is neatly summed up by Everett and colleagues in *Nature,* June 13, 1970: "It is premature to reach final conclusions as to the origin of the phenomena of anomalous water, and theoretical calculations of possible structures of water polymers are largely irrelevant until the need for a description in terms of polywater is called for. An equally, if not more important theoretical problem will be that of explaining why such polywater should exist in 'frozen equilibrium' with ordinary water."

This was sound advice from one well versed in thermodynamics and surface chemistry, but it was lost on those whose main concern was the establishment of a new molecular structure. To them, once Deryagin's methodology had been tested out and found to work, the main challenge lay in spectroscopic studies, which would provide an insight into the fundamental differences between ordinary water and polywater. The early experiments of Lippincott and his colleagues were symptomatic of this type of approach. Soon yet another group—the theoreticians—became involved; their aim was to test the speculations of the spectroscopists by applying the fundamental theories of atomic and molecular structure. Somewhat belatedly, the analytical chemists also joined in and brought to bear a battery of ever more refined experimental techniques to establish the true chemical composition of polywater. The suspicion of impurities was ever present, and to the analysts the prime question was whether impurities (if present) might not be somehow responsible for the strange properties of polywater. With never more than a few micrograms available, ingenious analytical methods had to be devised if definitive and quantitative results were to be obtained that would themselves stand up to scrutiny. The instrumentation required for much of this work was of the most advanced kind, only available at a few centers in the United States. In the end the analysts carried the day, because they showed, to their own and most other scientists' satisfaction, that polywater really contained little water, and that its composition was quite variable.

In 1969 and early 1970 most of the debate centered on the questions of what might be the molecular structure of a liquid, allegedly composed solely of oxygen and hydrogen, that exhibited the physical properties of polywater. It might be said that the scientists got their priorities mixed up, and that the first issue to be settled should have been whether the liquid was indeed composed of oxygen and hydrogen only. Sir Arthur Conan Doyle has warned us through the words of Sherlock Holmes that ''it is a capital mistake to theorize before one has data.'' But such is the nature of scientific enquiry: It is much more exciting to produce a new theory of chemical bonding than to establish that the remarkable behavior of a novel substance arises from the presence of x percent sodium chloride. Most scientists are quite ignorant of the extreme precautions and laborious procedures that must be followed by the chemical analyst trying to establish the presence or absence of an unknown quantity of one or more unknown substances which *may* be present in minute fractions of a percent in a sample of microscopic dimensions. Such issues hardly raise a yawn among the theoreticians, the aristocrats of the scientific hierarchy.

Let us take a closer look at the interplay between the spectroscopist and the theoretician, as demonstrated during the polywater debate. I have already made reference to the concepts of ''water structure,'' ''bound water,'' and water ''modified'' in some ill-defined way by its proximity to solid surfaces; now we have to enquire more closely into what makes water different from other small molecules, and how polywater might be yet different from ordinary water. In order to do this, we must get right inside the molecule and look at the way in which the electrons are distributed, the directions of the bonds between molecules, and the strengths of such bonds. Spectroscopy provides the most direct experimental tools—to put it the other way, spectroscopy is the language of the molecule. We measure the intensity of the radiation absorbed or emitted by the molecules over some particular range of the frequency spectrum, and in principle such measurements as shown in figure 7 contain all the information about the distribution of the electrons within the molecule, about strong chemical bonds between the atoms, and about the much weaker bonds between molecules. The trouble is that for all but the very simple chemical compounds the appearance of the spectra can be quite complex and their interpretation ambiguous. The molecule is telling us all about itself, but we cannot understand the language very well. It is here that the theoretician tries to help out by examining the problem in reverse. Starting with a set of nuclei and electrons, and using certain fundamental rules, he *calculates* the manner in which the electrons are distributed in space, the resulting shape of the molecule, and in favorable cases even the way in which several molecules might interact. These calculations become ever more approximate and uncertain as the total number of electrons to be disposed of increases. The water molecule is composed of three atoms (two hydrogens and one oxygen), and therefore has three nuclei. The oxygen atom has eight electrons and the hydrogens have one each. The calculations thus involve three nuclei and ten electrons, which is within the bounds of the possible.

The energies of chemical bonds, the bonds that hold together the various atoms within a molecule, are very large compared with the energies of bonds *between* molecules, that is, the weak interactions that are responsible for "structure" in liquids. In the case of water the situation is even more complex because of the overriding importance of the "hydrogen bond." Figure 2a shows the architecture of the H_2O molecule. Although it is electrically neutral, the electron distribution is such that the oxygen side carries a small negative charge and the two hydrogen atoms possess small, balancing positive charges. There is therefore a tendency for two neighboring water molecules to line up so that the negative oxygen end of one is adjacent to a positive hydrogen end of its neighbor. This arrangement provides a so-called hydrogen bond, which has a strength that corresponds to about 5% of that of an ordinary chemical bond holding hydrogen and oxygen together *within* the molecule but is enough to make the water molecules "sticky."

Because of this stickiness, water molecules come together more frequently and break apart less frequently than they otherwise would. This phenomenon accounts for the high boiling point of water, because enough energy has to be provided to overcome the stickiness. On the other hand, as the temperature is lowered to $0°C$, hydrogen bonding becomes more prevalent and the water molecules lock into place to form ice, as shown in figure 2b. Closer inspection of the ice structure reveals that each oxygen atom has two hydrogen atoms close to it but is hydrogen-bonded to its neighboring oxygen atoms through two more hydrogen atoms farther away which "belong" to another oxygen atom. Hydrogen bonds are not peculiar to water but occur in many different chemical contexts, wherever a hydrogen atom finds itself between two other atoms that carry negative charges or, as they are called in the trade, unpaired electrons. The hydrogen bond is one of the most important factors responsible for the maintenance of complex biological molecular structures, such as proteins and DNA. However, hydrogen bonding in water is unique, because every H_2O molecule can take part in *four* such bonds (see figure 2b) directed toward the corners of a regular tetrahedron. This can give rise to three-dimensional structures which distinguish water from other liquids.

The fact that the concept of the hydrogen bond can explain the physical properties of water in the solid and liquid states is reflected in the spectra of liquid water and ice. These are exceedingly complex in appearance, and even now debate continues about their correct interpretation in terms of the detailed molecular features of water in the bulk. Lippincott's startling new spectra (figure 7), taken together with Deryagin's findings, clearly meant that in polywater the H_2O molecules must be subject to very different types of interactions. Lippincott drew the natural conclusion that in polywater he was dealing with a completely different substance, and to him the evidence favored a polymeric structure. This left open a most important question: What sort of bonds between oxygen and hydrogen could promote such stable polymers? It had to be something akin to a chemical bond—that is, something much stronger than the familiar hydrogen bond.

At this point the theoreticians became intrigued. Leland C. Allen of Princeton University, aided by Peter Kollman, then a graduate student, addressed this problem. Lest it be thought that the theoretician operates from the tranquility of his office and requires nothing but pencil and paper, perhaps a word about the *modus operandi* of theoreticians is in order. Nowadays theoretical science resembles experimental science in that it also requires expensive hardware—the digital computer. The calculations performed by Allen and Kollman on polywater are based on some of the very fundamental concepts of science embodied in the discipline of quantum mechanics, which sets out the quantitative relationships and the rules that govern atomic and molecular structure. Unfortunately, a lack of the necessary mathematical knowledge prevents us from being able to solve some of the basic equations. In addition, not all equations have "solutions" in the closed form. Approximate solutions can be achieved by so-called numerical techniques, and these involve a series of iterative operations of such complexity that an advanced computer is required. The availability of such computers has given a boost to theoretical studies of molecular structure and bonding. It has also led to a generation of scientists (including social scientists) who, on occasion, substitute computer power for insight and originality.

The value of computer-based calculations is proportional to the amount of thought that goes into the experiments and the programing, and also to the amount of cash that can be spent on calculations. In other words, it is quite possible to cut corners by making certain simplifying assumptions, but the inevitable result is a greater degree of uncertainty in the significance and accuracy of the answer produced by the computer. Essentially all scientific research rests on assumptions of one kind or another, and the practitioner must always be aware of the nature and likely consequences of such assumptions and of the possible results of corner cutting. Nowhere is this more important than in the exploitation of high-speed computers in the solution of difficult theoretical problems.

Leland Allen was quite familiar with the pitfalls of chemical research by computer, and his first-hand experience with hydrogen bond theory provided a natural entry to an involvement with polywater. Indeed, Lippincott's polywater spectrum should have excited all but the most phlegmatic theoretical chemist. As things turned out, Allen's wholehearted commitment to polywater made him the subject of much criticism and abuse—most of it quite undeserved, and some of it, regrettably, from elder statesmen of science. Allen and Kollman set out to investigate possible structures that might be consistent with quantum mechanics, given what was known about H_2O, and that could be reconciled with the physical behavior of polywater and in particular with the spectrum recently published by Lippincott. By going back to first principles, they derived a model that had some of the features of Lippincott's postulated polymer structures but went beyond it. This led them to suggest that the water molecules were bonded into symmetric cyclic cages, and they coined a new description: "cyclimetric water." Figure 8 shows the essential features of Lippincott's polymeric structures and Allen's hexagonal cages.

There are two highly unorthodox features to "cyclimetric water." Following Lippincott's suggestion of "symmetric" hydrogen bonds—with the hydrogen atoms situated halfway between the neighboring oxygen atoms—Allen and Kollman further postulated that each oxygen atom could participate in *four* such bonds. This then enabled the hexagons to be stacked in different ways, as shown in figure 8 (bottom), each of which was quite consistent with the polywater spectrum. The stacked hexagonal assemblies were also thought to be compatible with the observed high density and viscosity of polywater.

One very surprising outcome of the extensive calculations was that the internal energy of cyclimetric water turned out to be very similar to that of ordinary water. The internal energy of a liquid is a measure of the attractive forces between the molecules. By conventional thermodynamic reasoning Allen and Kollman could now explain why the spontaneous generation of polywater should be so unfavorable and why, therefore, the substance was not commonly found in nature. Water and polywater were thought to be analogous to the two forms in which carbon exists in nature, diamond and graphite. In this analogy the diamond structure closely resembles that of a well-known form of ice, and graphite, made up of layers of planar hexagons of carbon, is the analog of cyclimetric water. The knowledge that many organic molecules (such as acetaldehyde and carbon disulfide) can exist both in the monomeric and the polymeric state lent further weight to the hypothesis.

As to the synthesis of polywater, Allen and Kollman thought that this was catalyzed in some highly specific fashion by the quartz or glass surface. It was postulated that on moving away from the solid surface there would then be "a smooth

"Anomalous water"
(Bellamy and Lippincott, May 1969)

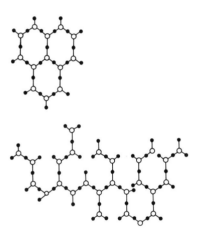

"Polywater"
(Lippincott et al., June 1969)

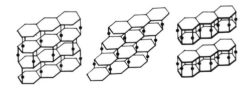

"Cyclimetric water"
(Allen and Kollman, March 1970)

Figure 8
Molecular structures proposed for polywater to account for its observed properties.

transition from an initial conventional hydrogen bonding through asymmetric cyclic hexamers, to the final three-dimensional lattice.'' The high stability ascribed to the six-atom ring made it unlikely that additional water molecules would condense on polywater once it was removed from the solid surface. The theory of Allen and Kollman was duly published in *Science* on March 13, 1970. Although experts can no doubt pick holes in the techniques, in thoroughness and thoughtfulness the paper compares favorably with anything that had previously been written on the clustering of water molecules or polywater. The conclusions bear the hallmark of a sustained intellectual effort, something that can hardly be said for the conclusions drawn by Lippincott and his colleagues nine months earlier.

Allen's theoretical investigations were followed by several similar attempts, most of them quite superficial by comparison, to derive for polywater a credible structure fully consistent with the physical properties so far established. The one feature of cyclimetric water that worried many scientists was the hypothetical transition of the familiar asymmetric hydrogen bonds $(O{-}H \cdots O)$ of ordinary water into symmetric bonds of a new type $(O{-}H{-}O)$. Chester O'Konski of the University of California tried to tackle this problem by computational techniques similar to those used by Allen and Kollman, but reached the conclusion that stable polymers of water ''probably'' could not exist. Another problem that might have been recognized earlier was the supposed stability of the hexagonal polywater rings. There was really no precedent for this; on the contrary, the rapid movement (known as ''tunneling'') of hydrogen atoms along the line of the hydrogen bond is a well-known phenomenon.

Peter Kollman recalls that, with the unavoidable assumptions and approximations inherent in complex quantum-mechanical calculations, the techniques employed were so crude that ''one could weight the evidence to either support or not support the existence of polywater.'' ''Early on,'' says

Kollman, ''we chose to weight the positive evidence more, since one could construct such a seemingly self-consistent picture of the phenomena.'' It was courageous (foolhardy, according to some) to try and tackle by quantum-mechanical methods a system containing so many electrons. On the other hand, if theoretical approaches were ever to prove themselves important to chemistry, then according to Leland Allen ''they had to be carried out *while a new area was being developed,* not after the answer was known experimentally, where theory would be acting simply to rationalize matters. I believed that theory should be used as a *predictive* tool. Anomalous water was an ideal case, because the problem was clearly important and the answer was clearly unknown.''

Not everyone involved took the polywater issue as seriously as did Leland Allen. Earlier on, Jerry Donohue of the University of Pennsylvania had published his own suggestions for the structure of polywater, based on clusters of fourteen water molecules. Although his contribution took up a page and a half of *Science* (November 21, 1969), it was, he said seven years later, ''concocted one evening while watching TV, between commercials, with tongue in cheek.'' Even so, at the time it was taken seriously enough to be criticized in scholarly terms by one of the senior members of the scientific establishment. It was also supported by funds from the Advanced Research Projects Agency of the Office of the Secretary of Defense. If nothing else, this anecdote illustrates how difficult it was at the time to judge what and whom to take seriously, even in the pages of reputable journals.

A final note on the theoreticians' attempts to assign a structure to polywater: Not long after the publication of their first paper, Allen and Kollman reported that, on the basis of further, more accurate calculations, they were now of the opinion that the proposed rings of cyclimetric water did in fact have a *higher* energy than six separate water molecules, and did not therefore constitute a credible alternative. In any case, the asymmetrically bonded rings, akin to real water, had a *lower* energy than the rings with the hypothetical symmetric bonds. This basic discrepancy between the theoretical predictions and basic thermodynamics as applied to the water-to-polywater transformation cast doubt on the validity and usefulness of the quantum-mechanical calculations. Of course, none of this invalidated the spectroscopic evidence or the possibility that polywater was indeed a polymer of water that was bonded in some as yet unknown way.

The matter did not end there. Over the next two years further reports appeared, mainly in *Nature* and *Science,* proposing ever more fanciful structures and accompanied by categorical statements and outrageous speculation. It would be unfair to claim that the following is typical of what went on at the time, but it did appear in *Nature* (August 22, 1970): "The anomalous water is therefore formed as six-membered rings . . . , hydrogen bonded to each other in a close-packed diamond-like arrangement. The O-to-O distance is taken as 0.24 nm and the hydrogen is centred between the two oxygens. Each oxygen is in turn surrounded by four hydrogen." The author then went on to speculate on the mechanism of formation of the anomalous water on the quartz surface, and ended with these statements: "One may certainly find the same mechanism in clay, soil, tissue and cells. It may also solve the question of why water can reach the top of a very tall tree." When it was pointed out to the author of this article (also through the pages of *Nature*) that his structural models for polywater were in fact nothing but scaled-down versions of two very common forms of ordinary ice, this elicited the reply (once again in *Nature*) "Why not?"

The contributions of theoretical approaches to the solution of the polywater mystery were disappointing even when rationally conceived and expertly carried through. The computing uncertainties are just too large, or the energy differences between alternative molecular structures too small, for such techniques to have any predictive value. This in no way invalidates Leland Allen's suggestion that for a theory to be worthwhile it must be able to predict the results of experiments yet to be performed. In the case of polywater, theoretical prediction proved to be impossible. This left the issue to be settled finally by the experimentalists—in particular, the analytical chemists, who perhaps should have been the first group to be alerted.

"For instance" is not proof.

Yiddish proverb

6

All Aboard the Bandwagon

The brief history of polywater reached its zenith in 1970. Much of the work that had been initiated directly as a result of Deryagin's various lectures or indirectly through the publicity by the Office of Naval Research came to fruition and found its way into the various scientific journals during that year. A major and well-publicized meeting on the subject was scheduled, at which, it was hoped, learned discussions would lead to a resolution of the various paradoxes surrounding polywater.

Alas, these hopes were not to be fulfilled. The daily press continued to take a lively interest, and as time went on the debate in the correspondence columns of the popular scientific weeklies became increasingly strident. It was difficult to remain objective, because the general atmosphere was one of polarization: One had to be a believer or a rejectionist. In the meantime Deryagin and his large team of assistants continued their investigations, apparently quite unperturbed by any suggestions that the strange properties of polywater might be due to contaminating materials.

In addition to the scientists on the firing line, there appeared an ever-growing flock of fellow travelers who sifted the evidence and wrote reviews, commentaries, and letters to editors in a variety of publications. Often it was hard to understand what prompted certain journals to publish these review articles, except that the subject was making headlines. At other times the choice of authors seemed curious to the initiated; among them were personalities well known in their own fields but relatively ignorant of the subject under discussion.

Some of these derivative articles seemed intended more to show off the authors' wit and erudition than to help those who were trying hard to unravel the contradictions surrounding polywater. For facetiousness it is hard to surpass the review in the *Journal of Inorganic and Nuclear Chemistry,* which, while claiming to examine current views critically, was really nothing but a six-page hodgepodge of quotations—many out of context—of some of the more eccentric claims that had appeared in the literature. This account, peppered with wit and erudite bon mots, and entitled ''Water: How Anomalous Can It Get?,'' contributed nothing of substance to the debate.

Along with the increase in scientific effort went an increase in speculation, in particular about the possibilities of exploiting polywater commercially and about the possible importance of polywater in life processes. The *Washington Post* of February 22, 1970, reported the hope that polywater would explain ''many perplexing questions in nature.'' For instance, polywater-type properties might be responsible for preserving winter wheat seeds from freezing and enabling insects to survive subzero temperatures during winter, and polywater research might even show botanists how water can reach the top branches of giant redwood trees.

One of the more curious sidelights of the polywater controversy was the suggestion that Venus might be a polywater planet. F. J. Donahoe, who had in 1969 warned society about the potential hazards of polywater research, now argued that the chemical evolution of Earth and Venus had diverged because ''polywater formed on the surface of Venus.'' He predicted that a detailed comparison of the reflection spectra of polywater with the spectra of the yellow-haze layer of Venus might well give positive indication of the presence of polywater. Donahoe's argument was carefully written and made interesting reading, although as a nonspecialist in planetary chemistry I had no way of knowing whether he had kept within the permitted boundaries of scientific speculation. On a more mundane plane, new spectroscopic work reported from Princeton University appeared to vindicate the pioneering work of Lippincott, who at that time was already claiming to have prepared polymerized forms of propyl alcohol and acetic acid. It seemed certain that the fluid now widely known as polywater could be readily prepared and that it indeed had the properties claimed for it by Deryagin.

The nagging doubts that these strange properties might be due more to impurities than to the molecular rearrangement of H_2O molecules were first given substance in a detailed report in *Science* of March 27, 1970, which described a chemical analysis of polywater carried out by a combination of very refined techniques at the Bell Telephone Laboratories and the University of Southern California. The senior author, Dennis L. Rousseau, became the chief protagonist of the "impurity lobby" that finally won the day. What surprised the scientific world in this report was the claim that polywater contained a whole range of inorganic substances but hardly any silicon. Up to that time it had been supposed, with reason, that any impurities present would originate from the material of the capillary tubes in which the polywater had been grown. Now Rousseau claimed that polywater contained from 20 to 60 percent sodium, and also potassium, chloride, sulfate, and traces of several more elements, even carbon— very little water. The disturbing thing was that the substance analyzed by Rousseau had been prepared by the standard methods and had all the properties characteristic of polywater.

The *New York Times* of April 2 carried the following headline (across all eight columns): "Researchers Cast Doubt on Finding That Water Can Be Converted to a Dense, Vaseline-like Form." After summarizing Rousseau's results the reporter concluded that "despite the new findings, a check of eminent scientists who believe in the existence of polywater showed no weakening of their conviction." Then followed the news of the first international conference on the subject, to be held under the auspices of the American Chemical Society at Lehigh University in Bethlehem, Pennsylvania. Characteristically, the article ended with the words: "The final showdown is expected at the June conference.'' By now many scientists were committed to such an extent that a showdown seemed inevitable.

Leland Allen, who at that time was right at the center of the controversy, doubted whether anything much would be achieved. In a letter to Walter Sullivan, science editor of the *New York Times,* a few days before the Lehigh conference, Allen wrote ''. . . it is now clear that the work to be reported at the Lehigh symposium will *not* resolve the problem whether anomalous water exists.'' The reasons were that insufficient amounts of the material were available and that there had been a lack of interchange of samples and measurements among the laboratories. The letter ended with the statement that anomalous water was ''potentially the most radically new chemical phenomenon in the last 50 years.'' Sullivan evidently agreed, and the *Times* gave the symposium a good deal of publicity.

Just prior to the Lehigh meeting there had been a further spate of polywater articles. Hardly a week went by without at least one account in *Science* or *Nature.* On April 3 *Science* reported attempts (unsuccessful) by Xerox Corporation scientists to generate polywater during an electric discharge in a gas. On April 11 *Nature* published a note by an Israeli scientist who, on the basis of theoretical considerations, had reasons to doubt the existence of a polymerized form of water. Not to be outdone, *Science* came back on April 24 with speculations from the General Electric Laboratories about the possible hydrogen-bond energy in polywater. Scientists at the North Dakota State University described anomalous water found in a common enzyme preparation. And so it continued. At the same time Deryagin summarized his work yet again, this time in the prestigious *Scientific American.* A caution was sounded by Nigel Calder, then science correspondent for the *New Statesman,* a well-known British left-wing weekly. In an article headed ''Polywater under the Bridge'' he reviewed the evidence and placed polywater in its historical perspective. He ended with a warning: "A final moral is one which is hard on would-be Einsteins. When experiment and a long-standing theory disagree, it is always much more likely that the experiment is wrong." This was sound advice, but in the grand tradition of sound advice it went unheeded.

On the eve of the 44th National Colloid Symposium, the supporters and opponents of Deryagin assembled at Lehigh. The Allentown *Morning Call* of the day carried two stories on polywater. One was the by-now-familiar rehash of polywater's physical properties and about its possible applications (which had increased in number and eccentricity; they now included the use of polywater as a preservative in blood banks). The other article mentioned that three hundred scientists were to attend the symposium, which would be opened by Professor Deryagin, the guest of honor.

The proceedings of the Lehigh symposium were eventually published, unfortunately almost a year after the event took place. As it turned out, Leland Allen had been right; the meeting did little to resolve the doubts and problems, and there was no showdown, although some of yesterday's convinced believers now openly expressed doubts. The questions posed by impurities and irreproducibilities in the preparation procedure could not be easily brushed aside, although Deryagin's line was unchanged: The presence of contaminants was evidence of careless work; his own samples were clean or only contained minute amounts of foreign materials which could in no way account for the measured physical properties of the modified water. This view gained support from some contributors to the symposium. In contradiction, Rousseau presented his findings, under the heading "An Alternative Explanation for Polywater," and most of the participants at the meeting could not help but be impressed by the care taken in his work and by the manner of his exposition.

The most damaging evidence to the cause of polywater came from none other than Ellis Lippincott. He told the meeting that there was now doubt whether the previously published spectrum (which had started the furor in the United States) was in fact indicative of a new, polymerized form of water. Lippincott now found that not only could he reproduce this spectrum by the use of synthetic mixtures of various organic acids, but, more important, he could no longer reproduce the spectrum when such materials were rigorously excluded. The other major blow to the believers in polywater came from Leland Allen. Allen now stated that, as a result of more detailed calculations, he was satisfied that the new cyclic, symmetric hydrogen bond was after all considerably *less* stable than the conventional asymmetric form normally found in water. He therefore concluded that ''we do not now believe in the existence of anomalous water.''

In yet another contribution to the symposium, entitled ''What Can Theory Say About the Existence and Properties of Anomalous Water?,'' Allen and Kollman eloquently defended the role played by theoreticians in physical research generally, using the issue of polywater as a specific example. The philosophy is set out in the opening paragraph: ''In sharp contrast to those performing chemical experiments, the theoretician is frequently called upon to justify his participation and to state explicitly what he expects to result from his labors.'' This is an interesting thesis, and one that in other contexts might have some validity. As regards polywater, however, there was a vague feeling that the theoreticians had been in too much of a hurry to become involved, on the basis of the very slender evidence provided by a single spectrum. Ten years later Allen was to regret the timing of the polywater episode, in that, had it all occurred just five years later, the powerful computers then available could have checked out the feasibility of speculative new electronic structures and molecular models ''the same afternoon that one first read the article.''

During and immediately after the Lehigh meeting the opinion was gaining ground that the involvement of the theoreticians had been premature, and that time and money had been misapplied to pursuits that were as yet of doubtful value. Allen's statement implies that experimentalists, as distinct from theoreticians, can spend money without having to give an account of their activities. This is hardly so; on occasions it costs more in time and effort to provide the required justification than it does to carry out the actual experiments. To most practicing scientists—theoreticians and experimentalists alike—Allen's sentiments must strike a familiar chord.

Looking back through the 1971 issue of the *Journal of Colloid and Interface Science,* which contains the collected papers delivered at the polywater symposium, one is struck by the lack of new material, that is, material that had not either been published in some form before the symposium or been "leaked" to the press. Certainly the discussions were polite and formal, but the symposium did little to clear up the outstanding puzzles. Apart from the professed converts Lippincott and Allen and those very few, like Rousseau, who were quite explicit in their commitment, the majority of participants remained where they had been before: sitting firmly on the fence. The contributed papers make heavy, though unrewarding, reading. The articles contain numerous unhelpful statements, such as these:

The presence of salt in anomalous water does not mean that there is no Water II present.
. . . we feel that X-ray diffraction, in addition to being a valuable means of identifying impurities in anomalous water, may also give much insight into the structure of a polymer of water if such a polymer exists.
The description of water-II as a polymeric system of H_2O with unusually strong, three-centered hydrogen bonds is not supported by our quantum mechanical calculations. However, we cannot at this time exclude the possibility of some stable "polymeric water" existing in a structure not considered in this paper.

Though the scientific dividend of the Lehigh Symposium was meager, the meeting received its share of newspaper publicity. Apart from its predictions of "public hanging," the *New*

York Times (June 24 and 28, 1970) described a press conference called by the American Chemical Society at which a poll was taken of the participants to the symposium. Those who felt 60 percent sure that polywater existed were asked to raise their hands. One or two hands went up. Similarly, only one or two were convinced that polywater did not exist. An almost unanimous vote was given in favor of the proposition that the question remained open. The *Times* also informed its readers of the impressive list of U.S. government agencies supporting research on polywater. Apart from the Office of Naval Research these included the Army's Cold Regions Research and Engineering Laboratory, the Advanced Projects Agency of the Defense Department, the National Science Foundation, the National Bureau of Standards, the Office of Saline Water of the Interior Department, and more. Professor Deryagin told the *Times* he had recently been authorized to recruit fifteen more members for his research team, perhaps to counter an emerging "polywater gap."

Headlines of the time reflect the newsworthiness of polywater:

Cette eau qui n'en est pas
Science et Vie (Paris), September 1970

Weird Water: Debate Over Mysterious Fluid Splits Scientific Worlds: Polywater has bright future—or does it? Critics charge it's only impure tapwater.
Wall Street Journal, July 21, 1970

Can He [Deryagin] Prove It's Water?
Chemical Week, July 1, 1970

Scientists Growing Wet, Creepy Water
Washington Post, February 22, 1970

Throwing Cold Water on Polywater
New Scientist, July 16, 1970

There were many other such headlines. Some stories translated into simpler language the reports in the scientific jour-

nals; others reported interviews with polywater personalities.
Some attempted a serious analysis of the puzzles posed by
polywater; others accented the "human interest" aspects of
the controversy. Despite the convictions expressed by supporters of polywater in the scientific articles, it seems that
nagging doubts persisted in the minds of some. In an interview with the *Washington Post,* Ellis Lippincott told the story
of his involvement with polywater. If not the father, then at
least the uncle of polywater, he confessed to the reporter that
"even now it's so strange that we keep wondering what's
the big goof we've been making. Is there something subtle
that's fooling us? We don't think so."

At this stage Robert L. Davis of Purdue University entered
the debate, not through the medium of a scientific journal
but via a press interview in *Chemical & Engineering News.* He
had apparently been working in collaboration with Dennis
Rousseau, although his name does not appear on any bona
fide scientific publication until 1971. Davis claimed that he
had obtained identical infrared spectra for polywater prepared from ordinary and from heavy water. As discussed earlier, the infrared spectrum sensitively reflects the details of
chemical bonds (in the case of water, the bond between oxygen and hydrogen). The hydrogen atom (called deuterium) in
heavy water is twice as heavy as the normal hydrogen atom,
and this difference in mass causes the spectrum of heavy water to be quite different from that of ordinary water. The findings of Davis, if accurate, proved quite conclusively that the
previously reported spectra, whatever their origin, could not
possibly have had anything to do with water. Davis also suggested that polywater contained impurities of an organic
character, and we were to hear more about this later in the
year, again through the medium of a press interview.

The remainder of 1970 saw a continued increase in the number of reports on polywater, mostly in *Nature* and *Science* but
also in such publications as the *Journal of Physical Chemistry*
and the *Journal of Chemical Physics,* where contributions are
thought to be rather more strictly reviewed than in the week-

lies. Thus, the impurities hypothesis of Rousseau gained powerful support from the extensive analytical work of two scientists at the Los Alamos Laboratory, which prompted the *New Scientist* to predict that "the stage is now set for a rebuttal from the polywater protagonists. Unless they come up with very convincing evidence, it is most likely that polywater will end up in the wings."

On July 4, 1970, polywater got a foothold in Australia when two scientists at the Sydney Government Laboratories for Food Preservation reported the discovery of a "new" polywater, one that could be grown on surfaces that did not contain silica and that could not therefore be contaminated by silica. The two Australians ended their report in *Nature* by stating "There seems no reason why a family of similar polywaters should not exist." In the light of what was then already known, the methods used for the preparation of this particular sample of polywater should have been highly suspect: The water vapor was condensed in the presence of an aqueous solution of potassium sulfate, whereas by now one of the ground rules was that inorganic salts must be rigorously excluded. However, once again a newspaper (the Sydney *Morning Herald*) published the story complete with interviews and speculations: "Polywater could bind the heat-resistant bacterial spores found in food. . . . The presence of polywater possibly explained the fact that even freeze-drying cannot remove every last drop of water from organic material. . . . Minute amounts of polywater are probably quite common in nature."

A rebuttal quickly came from William Mansfield, also an Australian government chemist, who, again in the Sydney *Morning Herald,* pointed out that the arguments in favor of the existence of polywater depended on the absence of impurities, but that this was impossible to prove. He also made a disclosure that was to achieve some notoriety, namely that he had been able to reproduce the infrared spectrum ascribed by Lippincott to polywater by using "a sweat substance as an impurity." Although no details were given, subsequent de-

velopments suggest that this may have been lactic acid, a
major constituent of perspiration.

Mansfield died in 1974, but a former colleague later wrote to
me that his experiences with the scientific journals had been
"somewhat unhappy." On several occasions his manuscripts
were rejected by *Science* and *Nature,* which later published
similar results by different authors. This type of complaint
can frequently be dismissed on the grounds of undue sensitivity on the part of scientists or even mild forms of persecution mania. However, Mansfield actually received a letter of
apology from an editor of *Science,* so his complaints may
well have had some substance.

This arbitrary treatment at the hands of editors—especially
those of the journals *Nature* and *Science*—is a recurring
theme in my correspondence with scientists who were at that
time involved in the polywater controversy. One can only assume that in the heat of the moment the review system had
suffered a temporary short circuit, or that in some instances
editors declined to submit articles for review and exercised
their prerogative to accept or reject. This must have led to
rough justice in the eyes of the authors. Nevertheless, one
can question ten years later whether the offerings on polywater that did appear in these prestigious journals are really
those for which researchers would like to be remembered by
future generations of scientists.

All the while the debate continued, with the weight of evidence in favor of impurities increasing all the time. It is rare
that a purely scientific issue receives the amount of publicity
that was lavished on polywater. A three-page article on the
subject in *Chemical & Engineering News* (July 13, 1970)
contains photographs of no fewer than five polywater
scientists.

Obviously polywater must have been considered as something of potential economic importance, since even the *Wall
Street Journal* printed an extensive summary of the Lehigh

symposium. It was more sensational than anything that could be found in the *New York Times:*

A few years from now living room furniture may be made out of water. The antifreeze in cars may be water. And overcoats may be rainproofed with water. These aren't fantasies if scientists succeed in getting out of the test tube what may be an entirely new form of water.

I may be lacking in imagination, but on first reading this in 1970 I thought exactly the opposite: that these were indeed fantasies. The article quoted several of the polywater "personalities" of the day, including Deryagin, who again refuted suggestions that contamination was the key to polywater:
" 'I cannot be responsible for polywater not prepared by us,' he sniffed."

An interesting sidelight on the reaction to Deryagin's claims within the Soviet Union was provided by *Science News.* It appears that opinion among Russian scientists was as divided as among their Western counterparts. At a meeting of the Soviet Academy of Sciences, Aleksander N. Frumkin, the director of the Institute of Physical Chemistry and a chemist of worldwide renown, asked "How much modified water has been obtained in all?," to which Deryagin replied "About enough for fifteen reports." Despite the levity, concern was expressed that no convincing chemical analysis had yet been performed, but Deryagin brushed this criticism aside by saying that he had been more concerned with establishing the properties of the new substance. Boris V. Nekrasov countered that "discoveries must be explained on the basis of materialism rather than on the basis of mysticism." Another eminent physical chemist, Pyotr A. Rebinder, averred that Deryagin had made "the discovery of the century." This echoes the views of some American scientists who, after the Lehigh symposium, predicted that if the discovery of polywater could be validated Deryagin should be a heavy favorite for a Nobel Prize.

One short sentence in the *Science News* story refers to a chemical analysis that had earlier been performed on samples of polywater. Organic impurities had been identified, but this evidence was played down by Deryagin. It is abundantly clear, even from the short transcript of the meeting, that Russian scientists were mystified and felt that there were too many unanswered questions for a final verdict to be reached. A followup of the Russian analytical studies brought to light that Deryagin had indeed arranged for twenty-five samples of polywater to be analyzed, and this had been done as early as 1968 by Vladimir L. Tal'rose of the Institute of Chemical Physics of the Soviet Academy of Sciences. The results (not publicized in the West) were reported in the obscure Russian journal *Chemistry and Life,* which is not translated into English. Tal'rose found that all the specimens supplied to him contained organic matter ''in quantities comparable with the amount of modified water itself.'' The nature of the chemical compounds found suggested that they might have had human origins, for instance perspiration. No evidence of a water polymer could be found in the samples. Subsequent analyses performed on fresh samples later in 1968 showed that the impurity levels were very much reduced.

This information first became available to Western scientists in the *New York Times* of September 27, 1970: ''Scientist Says Mystery of Polywater Has Been Solved: Russian's test samples contained sweat.'' The story had its origin in a telephone interview with none other than Robert E. Davis of Purdue University. Unlike the earlier polywater stories in the *Times,* the Davis interview was handled by an inexperienced and certainly somewhat naive reporter who, for whatever reasons, did not continue for long as a *Times* science writer. The story left the impression that polywater had been reduced to sweat. This theme was elaborated in the correspondence columns of *Chemical & Engineering News* on the following day, September 28, 1970. Here Davis produced an English translation of the analytical studies by Tal'rose. This was helpful. What was not helpful was the general tone of Davis's letter: ''Like the geysers of Yellowstone, the polywater controversy bubbles and steams but still produces little water.'' It finished: ''Thus we must conclude that all polywater is polycrap and that the American scientists have been wasting their time studying this subject, unless, of course, it can be defined as a topic of water pollution and waste disposal. Thus our own work with that of Rousseau . . . was a complete, polluted waste. It is hoped that these comments will take all the sweat out of the poly debate.'' For obvious reasons this hope was not fulfilled.

The topical news magazines were very quick to pick up Davis's allegations. While *Le Monde* of October 14 was still publishing all the well-worn pre-Lehigh material—''Le mystère de l'eau anormale''—*Time* and *Der Spiegel* printed the Davis story. One of the headlines: *Zeit vertan* (time wasted). The *Time* article was suitably illustrated with a picture of Davis at work in his laboratory, collecting perspiration for his experiments by wringing out a T-shirt. We had now been treated to two Davis portraits but were still waiting to see his scientific contributions to the polywater debate in print.

The replies to Davis's letter in *Chemical & Engineering News* were not long in coming. Barry Brummer rightly wondered whether the polywater phenomenon was more interesting scientifically or sociologically, since scientists (and others) had taken extreme positions, often with little regard for the facts. He discounted the ''sweat theory'' but left open the possibility that dissolved glass components might be responsible for the polywater properties. He ended on a general note of admonishment: ''By all means let us argue hotly the merits of the various theories—the outcome happens to be important. However, it might be better to minimize the polemics and to concentrate on the facts.''

These sentiments were echoed by others, both at the time and in subsequent correspondence. It was pointed out that scientific capital can easily be made by flogging a dead horse; that is, by refuting what has already been shown to be faulty. Most of the polywater researchers were displeased (to put it mildly) with the manner in which Davis had intervened in the debate. In a sense it was regrettable that what had started out as objective scientific debate had within 18 months degenerated into a thinly disguised slanging match, carried on in the popular news media.

By the end of 1970 the majority of experts were of the opinion that water as a polymer might or might not exist, but that the eccentric properties of the substance known as polywater resulted from relatively high concentrations of foreign matter. The exact nature of the impurities was still a subject of debate, as was their origin. On the one hand there was the meticulously performed work of Rousseau, which had identified various mineral substances. On the other hand it seemed more reasonable to advance the hypothesis that the impurities originated from interaction between the condensing water vapor and the walls of the capillaries. Then again there was the suggestion of organic contaminants. Finally, there were reports that stopcock grease from the vacuum system could creep along the glass or quartz surfaces and contaminate the condensate in the capillaries. None of these propositions could explain Fedyakin's original observation of secondary liquid columns which grew *spontaneously* in *sealed* capillaries, but by then Fedyakin and all that had long been forgotten.

Evil communications corrupt good manners.

1 Corinthians 15:33

7

The Patterns of Communication

Before tracing the polywater chronicle to its inevitable end, let us survey the methods available to scientists for the dissemination of their results. These methods have become institutionalized over the past two centuries, and there is now a good deal of ritual and convention associated with the communication among scientists. We shall see that the manner in which polywater was handled by the scientific community was on many occasions at odds with currently accepted norms.

The communication networks used by scientists are by no means peculiar to them; they are common to most scholars. Indeed, some of the techniques have more recently been adopted by other professions whose primary function is hardly the quest for knowledge. The most formal method for communication is through the scientific journal. Since the inauguration of the *Philosophical Transactions* as the official publication of the Royal Society of London in 1753, the scientific literature has mushroomed. New journals continue to appear at an alarming rate, and the day seems to be approaching when all the newsprint produced in the world is turned over to the production of scientific journals. Although the number of practicing scientists has also increased dramatically, especially since the 1950s, it has not equaled the growth of scientific journals. Since there is apparently no lack of would-be authors, a possible inference is that the scientist of today is more productive than his counterpart of a generation ago—but more of this later. The exponential growth of scientific publications is an acute embarrassment to librarians, who must manage on fixed budgets, and to scientists, who must keep abreast with current developments.

Clearly, not all journals enjoy the same standing with members of the scientific community. Some of the old established publications, especially those produced by learned societies, enjoy a high prestige due mainly to the professional standards they jealously guard. Unfortunately, their production styles tend to be out of date, their reviewing process is so formalized and their backlog so great that the publication process is subject to inordinately long delays, and their marketing policies (if any) are hardly aggressive. At one time scholarly publication was the sole prerogative of the learned societies, but in recent years commercial publishing houses have entered the market in increasing numbers. Since the dissemination of scientific data is hardly their prime consideration, it must be assumed that scientific journals offer possibilities of lucrative diversification.

It is agreed among practicing scientists and among sociologists of science that the prime motive for publication is the establishment of priority, the staking of a claim. The anxiety for priority is nothing new in scientific research. Indeed, by the standards of Galileo, Newton, Leibniz, Cavendish, and Lavoisier, today's debates are conducted with the utmost of decorum. This overriding concern for priority is hardly compatible with the slow birth of a new theory, or with the painstaking experiments required for the verification of such a theory. Nor do production delays by the publisher expedite the establishment of priority. To circumvent such delays and also to enable researchers to stake a claim before a complete, formal paper can be prepared, a number of journals have appeared where the accent is on fast publication of short articles. The implication is that such short announcements will later be followed up with detailed reports in regular journals. So, for instance, the old established *Physical Review* produced the offshoot *Physical Review Letters* as a vehicle for fast publication.

The time interval between the receipt of a manuscript and its eventual publication in a regular journal may well be a year or even longer. Several of the papers presented at the Lehigh polywater symposium were received by the *Journal of Colloid and Interface Science* on July 20, 1970, and were finally published in the August 1971 issue. By comparison, a short communication on the nature of hydrogen bonds in polywater by Keiji Morokuma of the University of Rochester was received by *Chemical Physics Letters* (then in its fourth year) on October 9, 1969, and was published on December 1. There is thus a distinct incentive for the researcher to avail himself of the letter-type journal, and to write the full story later in the knowledge that his claim is safe. An analysis of the polywater publications shows an inordinately large number of such short articles over a very short time span. Atypically, only a handful of these were ever to be followed up by more substantial accounts.

No doubt, some of the responsibility for publication delays must go to the review procedure, which has become the quality-control system of scientific communication. An article published in a learned journal does not just represent the opinion of the author, but bears the imprimatur of the professional society responsible for the publication. The editor of such a journal has sent the manuscript to one or several referees who act as judges of the credibility, authenticity, value (in the sense of advancement of knowledge), correct acknowledgment of previous work, and taste. The history and vagaries of the refereeing system make fascinating reading. It may not be an ideal system of judgment, but it is the best we have and it usually operates well. Cases have been known where the innocent have been hanged, but the victim has often lived to see his work and reputation vindicated.

The statistics of acceptance and rejection show that for the physical sciences the rejection rate is low, about 20 to 30 percent of the papers submitted; this compares with 90 percent in some areas of the humanities and 50 to 80 percent in the social sciences. For obvious reasons, papers devoted primarily to the presentation of results and observations stand a good chance of acceptance. In fact, the policy is usually to publish unless there is a serious objection to the experimental techniques or the reliability of the results.

More often than not referees are drawn from the scientific elite, that is, from among those working in well-known and well-respected departments. For instance, a study of the refereeing policy adopted by *Physical Review* over many years has shown that two-thirds of the referee judgments came from just 17 of the major physics departments of American universities. This does not imply that the chosen referees were necessarily older men, but it does suggest that they benefited from the professional reputation enjoyed by the department. The acceptance rate also did not seem to depend on the age of the referee, but there was every sign that highly placed physicists had disproportionate power in relation to their numbers in deciding what would enter the pages of *Physical Review*. On the whole this policy has achieved good results, because the more experienced scientists often assist younger colleagues in the ''rehabilitation'' of papers that have previously been rejected.

Apart from assessing other aspects of a paper, the referees monitor the style and tone of the writing. The currently acceptable style is bland and stripped of all emotion. The work is presented in a historical, logical sequence, quite in contrast with the way in which discoveries actually come about. Comment on fellow scientists is expected to be soft-spoken and discreet—in contrast with eighteenth-century debates, such as the one between Cavendish, Watt, and Lavoisier, which turned quite harsh and ugly despite the acknowledged personal modesty of the protagonists. The polywater debate showed distinct signs of reverting to this long-discarded style of personal abuse, not because the priority of Fedyakin and Deryagin was ever challenged but because on both sides of the debate and among the bystanders there were those who would have impugned the motives and the professionalism of some of the protagonists. (Because most referees would have disallowed such comment, the debate was carried on in the popular scientific weeklies and the news media, which are not subject to refereeing procedures.)

It is also the task of the referee to ensure that priority is acknowledged and that the bibliography contains suitable references to previous work. Selectivity in citations to previous work provides the scientist with an additional political device. It is in most cases quite impracticable to provide a complete bibliography; a selection has to be made of prior work that bears some relevance to the present article. There are those in the scientific community who are prone to emphasize their own work at the expense of contributions made by their competitors. In extreme cases they make no reference at all to the publications of certain individuals, or, worse still, they only refer to such work in order to show up its weaknesses. The polywater literature provides a model of fair referencing. The battles were fought elsewhere: at congresses and via the scientific grapevine. We will consider these channels shortly.

In the early years—up to 1966—work was confined to the Soviet Union. Only ten scientific papers were published, but even they contained much that was repetitive. During 1966 and 1967 five reports from Britain made speculative references to anomalous water, without contributing anything of substance. American scientists first reported activity in 1968, but outstripped the Russians by 1969. In 1970 some 120 articles were published; the fact that as many as 25 fell within the category of "vulgarization" indicated a general interest in the subject. The patterns of publication are shown in figure 9. Several interesting features emerge. Research activity in America was confined to the short period 1969–1972. (Since there is usually at least a six-month lag in publication in technical journals, the period of actual research should be shifted backwards somewhat.) Even before American scientists got deeply involved, there had already been several references to anomalous water in the American technical and popular press and in government publications. Similarly, the

120

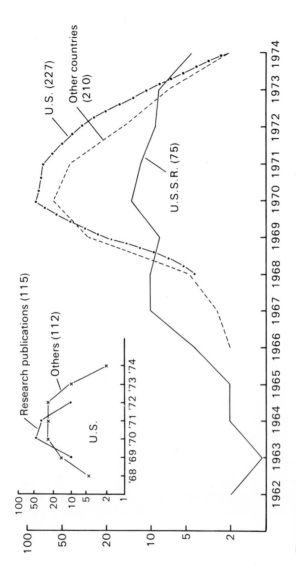

Figure 9
Polywater publication patterns for the period 1962—1974 (total number of publications in parentheses). Inset compares research publications and other articles (commentaries, reviews, and popularizations) published by U.S. authors.

aftermath (1973 onwards) also produced a considerable number of such references. A comparison of the Western and Soviet communication patterns illuminates the different ways in which scientific research is organized. The Russian polywater investigators never published more than nine articles on original work in any one year, and even after the remainder of the world had lost faith the Russians persisted with their studies of anomalous water. Even now they are still actively pursuing investigations into the nature of liquid films on solid substrates, and that is how it all started.

Another interesting and significant aspect of polywater reporting in the scientific journals is that the large majority of original research was published in the form of short communications in *Science* and *Nature,* few of which were subsequently followed up by more detailed reports in specialist journals. This testifies to the urgency with which the subject was treated. On the other hand, as Pieter Hoekstra points out, ''In retrospect it is evident that the desire to get material in press and thereby enter the mainstream of excitement affected sound judgment.'' Hoekstra makes this criticism in relation to his own contribution to the literature of the time, but it is applicable to much of the work that was submitted for publication. There must have been considerable pressure on the editors of the two leading scientific weeklies. There certainly was rough justice in the choice of papers for publication. Many of my correspondents remember occasions when their manuscripts were rejected, delayed by referees, or unjustly criticized. Walter Madigosky of the Naval Surface Weapons Center put it most graphically. Madigosky's spectroscopic studies had led him to the conviction that the polywater spectrum resembled that of sodium acetate rather than sodium lactate, a major component of sweat. (The connection between polywater and sweat had received much publicity in the popular press.) He writes: ''I tried to get a news release from my laboratory on my results but management decided not to commit itself but wait until my paper has been accepted for publication. I wrote up my results and sent it to

Science. At this point the editor of *Science* was up to his neck in polywater, thus he was somewhat hesitant to accept another paper, but finally consented to a shortened version.''

The number of full-length papers from Western countries, excepting those published as the proceedings of the 1970 Lehigh conference, did not exceed ten. Leland Allen has pointed out that one of his papers was the only polywater article ever to be published by the *Journal of the American Chemical Society.* The inference is that very few papers on the subject were of sufficiently high quality to survive the strict refereeing practiced by this journal. By no means did Allen have an easy passage with the referees. Behind their shroud of anonymity, referees attacked the authors rather than the work. One such report consisted of a set of symbols usually employed for swear words and vulgar expressions in comic strips, followed by the single word ''reject.'' Nevertheless, the editors treated this type of referee's report in the same bureaucratic manner usually accorded more rational comments. Allen's paper was finally accepted only because an eminent authority on quantum mechanics certified that there was nothing wrong with the techniques used.

The editors of *Science* and *Nature,* although they solicited referees' reports, were more likely to take chances by publishing borderline material. The dilemma that faces an editor of a topical science journal has been graphically described by Theodor Benfey, himself an editor of such a magazine (*Chemistry*). In his obituary of polywater he writes: ''New theories come seldom, and facts that fit theories are seldom interesting. 'Facts' that don't fit theories come a dime a dozen, and the best are fun to describe and discuss. The editor must try to separate fact from fiction and sloppy technique from brilliant instrumentation, and must distinguish between publicity seekers and those courageously presenting new information.'' By these criteria several blunders were made in the reporting of polywater research, but at least the debate went on.

Communication by means of scientific journals is laborious and slow. A more direct way of passing on new results and information is provided by the many congresses at which scientists meet. Congresses come in different sizes. The mammoth gatherings of the American Chemical Society are at one end of the spectrum. Here, under the umbrella of one parent organization, different specialist groups arrange their own symposia, so that at any one time up to a dozen sub-meetings may be in progress. More often than not these take the shape of a succession of so-called "short contributions" in which the author is given 10 to 15 minutes to present a paper and to answer any questions. One of the main reasons for this format of meeting is that in many cases travel and other expenses will not be reimbursed by the employer or the granting agency unless the applicant presents a paper. This places a premium on fast talking, as opposed to deep listening and learning. The audience has the hard tasks of assimilating what the author is telling them, remembering whatever seems important, and asking the right questions. After surviving two days of a "short communications" barrage, only those with above-average stamina come back for the third and successive days. The session chairman has the unenviable task of keeping the meeting to the printed timetable, because many members of the audience only show up for specific papers, after which they disappear and make their way to another session where a paper of interest is being presented. The chairman's job is not made any easier by the inevitable speaker who insists on showing some thirty transparencies within the ten minutes. This has led to the practice of members of the audience photographing the projected transparencies, because there is no time to take notes. One must on occasions wonder how much knowledge and information can actually be assimilated at such a congress. (Of course, there are periods when no lectures are scheduled, or when participants are too exhausted to face another day in a hot dark lecture theater, peering at slide projections which can on some occasions only be read from the

124

front three rows. During such periods one finds a friend or two and visits the bar or the coffee shop, where scientific debate flourishes.)

Another feature of large meetings is that the initiated seldom learn very much that is new to them. This was the case with polywater. Invariably, one or the other of the chief polywater spokesmen would deliver a set-piece lecture describing work that had already appeared in print weeks or months before. Though these talks (which were at best rewrites of identical lectures given at other meetings) might have bored the informed scientists, they did draw the press. Much of the publicity devoted to polywater originated with such lectures delivered at one or another large meeting.

Fortunately, there are other types of meetings with fewer participants and less ambitious programs. Participation may even be limited or by invitation only, as is the case with the Gordon Research Conferences. The lectures delivered at a meeting may subsequently be published. In some cases, such as the Faraday General Discussions, the papers are first refereed for originality, and those accepted are precirculated for every participant to read in advance, so that at the meeting the time is devoted exclusively to their discussion. Polywater made appearances at all types of meetings; it was even considered important enough to figure in *several* Gordon Research Conferences. (In some instances it was not easy to spot the connection between polywater and the subject of the conference, but polywater was news.)

By far the quickest and most effective way of passing on new results and information is via the grapevine. Every scientist belongs to a ''club'' whose members are active in the same general area, use the same jargon, publish in the same journals, attend the same meetings, and are readily identifiable as a homogeneous group. Members of the club invite each other to give seminars and circulate the most up-to-date results for comment and discussion. Such privileged information may be quoted in the technical journals, where it is usually referred to as "personal communication" and duly attributed to the originator.

Membership in a club and access to the grapevine also figure in the race for priority. A hot topic stimulates lively interaction, letters, telephone calls, and invitations to present seminars. Since members of a club are well acquainted, everybody knows who his competitors are. This is particularly true among "top" scientists. Lower down the hierarchy, and outside the club, one is not so sure of the identity and location of the competition. The only alternative, then, is to publish quickly to establish a claim. This may help with election to club membership.

Some scientists belong to more than one club, especially if their research interests lie on the border between several disciplines, or if their work appears to be of very wide interest or applicability. Anyone who wears the label "student of water structure and aqueous solutions" has full or honorary status in several different clubs. It is a widespread assumption that what little basic knowledge members of the "water club" possess should enable them to solve many of the problems in pharmaceutical chemistry, food processing, biochemistry, colloid science, several branches of medicine, and so on. The practical result is that members of the "water club" are frequently invited to give seminars or keynote lectures at meetings of other clubs. On these occasions they are treated as prophets, and it is brought home to them how much truth there is in the saying that in the kingdom of the blind the one-eyed man is king. Yet, were it not for such cross-fertilization, each group of specialists would be even more divorced from every other group.

When professional and financial pressures demand an ever greater degree of specialization, it is easy to lose sight of general scientific principles and progress. In this aspect the polywater affair was atypical, because it was in no sense the preserve of any one closed specialist group. On the contrary, it was a meeting ground for several very different disciplines; this made the debate very stimulating and competitive. The unusually large number of references to ''private'' or ''personal'' communications in the published articles reflected a lively debate by telephone and letter and at various informal meetings. In this sense, the communication system associated with polywater was healthy and positive.

It is doubtful whether many nonscientists ever glance at *Science* or *Nature,* let alone the more abstruse journals, which to a nonspecialist look forbidding. There exist many popular science journals, some of them organs of professional bodies, such as *Chemical & Engineering News* or *Chemistry in Britain.* Others are published as commercial ventures and are aimed at the well-informed lay reader, such as *New Scientist, La Recherche,* or *Scientific American.* Such periodicals commission well-known or recommended specialists to write articles. They also have staff writers who are familiar with current scientific issues and their political, economic, or sociological ramifications. Such writers are not always particular about technical detail (indeed, they sometimes make the scientist wince), but they are at least adept at describing complicated concepts in language that the interested lay reader can understand. This is something few scientists seem to be able or willing to do. The French expression for such popular technical journalism, *articles de vulgarisation,* is hard to translate but catches the flavor particularly well.

Should a technical issue ever be considered to have real news value, the mass media will provide the publicity. Such instances are relatively rare, and the favored topics tend to be alleged breakthroughs that may perhaps lead to a cure for cancer, or more exact prediction of earthquakes, or greatly improved crop yields. Such stories usually include interviews with, or quotes from the researchers concerned. The possibility of such exposure is almost irresistibly alluring to scientists. Polywater had more than a normal share of press attention, and this was a major contributing factor to the general souring of the atmosphere as time went on.

Indeed, the manner in which polywater research was reported in the various media and at conferences exhibited certain features which, if not unique, were certainly uncommon. Marcel Pierre Gingold has analyzed this pattern of communication in a perceptively written *article de vulgarisation* in *La Recherche*. Gingold believes that polywater, as an episode in science, was abnormal right from the beginning. He reminds us that according to Karl Popper scientific research proceeds in an orderly manner: A hypothesis is developed and then either refuted or modified on the basis of its match with experience; eventually, if it holds up, it becomes a theory and takes its place in the general framework of scientific knowledge. Alternatively, a discovery is produced in the laboratory, its true existence is established, and it must then be interpreted in terms of existing theories; otherwise the theories are put in doubt. In the case of polywater the existence of the material as a new form of water was never really established before people started constructing theoretical frameworks, postulating new types of hydrogen bonds, and engaging in detailed debates about the molecular structure of the new material. In other words, right from the beginning the real issue was not whether the observed phenomenon was real, but how it was to be interpreted. This rather unusual approach, coupled with the facts that the material was one of the commonest known to man and that scientific enquiry was relatively free during the late 1960s, was responsible for the uncommon patterns of communication.

If we look at the growth and decline of polywater publications, we can get a clear sense of the distinctive evolution of this episode. Most scientific developments show the pattern of activity and publication represented by curve I in figure 10. New discoveries or major new advances in a

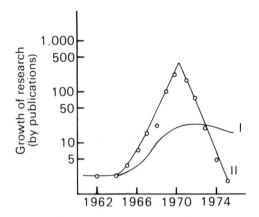

Figure 10
(I) Normal growth curve for a developing research area. (II) Growth curve for polywater research (corresponds to the course of an epidemic).

given area exhibit a steady growth, after an initial lag period during which the new subject has to achieve credibility and respectability. Eventually the growth curve flattens out and then gradually declines.

The three stages of development have been called the "pioneer," "bandwagon," and "rabbinic" stages. Often new ideas or concepts face stiff resistance from referees and indifference from the scientific community at large. Eventually acceptance follows, and a new discipline or subdiscipline may be born. Its practitioners will form their own "club," and their achievements may stimulate excitement and activity or even lead to the founding of a new journal. In the end, however, the law of diminishing returns takes its toll, and scientists begin to look elsewhere for new excitement. The discussions turn more and more to points of terminology and interpretation as the well of discovery runs dry. Such areas of science stand in danger of becoming baroque in style and content; they are no longer relevant to the mainstream of scientific development. In some circles it is even now being murmured that high-energy physics (as distinct from nuclear physics) has reached this stage of development.

The growth and decline of polywater activity looks entirely different from the norm. Bruce C. Bennion and Laurence A. Neuton of the University of Southern California have compared the growth of polywater research to an epidemic, and the resemblance is indeed striking. According to their analysis, all those scientists who had the necessary training and background for becoming active in polywater research are classified as "susceptible" and all those who published original work are referred to as "infected." Because of the universal importance of water in the natural and life sciences, Bennion and Neuton consider that a conservative estimate for susceptibles would be 100,000. Of these, 430 became infected, all of them directly or indirectly from the original source, Fedyakin and Deryagin. (My own reading of the situation, confirmed by most of my correspondents, is that Lippincott was mainly responsible for the outbreak of the epidemic. Had it not been for his widely publicized article in *Science,* describing the spectrum of anomalous water and coining the name polywater, the issue would not have caught the imagination of American scientists in the manner that it did. More likely, polywater would just have remained the preserve of the surface chemists and would have been disposed of in the normal way.)

A feature of real discoveries is that they are often made simultaneously, but independently, by several groups of workers. This was not so for polywater; the discovery was Fedyakin's only. This should have been enough to raise doubts about its credibility quite early on. If we now examine the rate of infection over the years (figure 10, curve II), we see that polywater became an "epidemic" in 1968, reaching its peak early in 1970. After that the rate of infection declined rapidly, and by 1971 scientists were recovering from the epidemic. Although 1975 still witnessed a small number of original publications on anomalous water, the scientific community had by that time become immune and the epidemic was over.

But had it really been an epidemic? Sherman Rabideau of the University of California, whose careful and painstaking analytical work helped to disprove the polywater hypothesis, recalls that the period 1969–1970 "was a time of high adventure—a new episode in the serial was undoubtedly forthcoming in the next issue of *Science* and *Nature.*" These sentiments have been echoed many times in my correspondence with former "infected" scientists. To be fair, there have been other views; Sefton Hamann of the Australian Government Laboratories in Melbourne recalls the period as "the darkest and most miserable in the history of physical chemistry." Be that as it may, the reporting of polywater certainly exhibited symptoms of a strange kind. We witnessed a multitude of short communications not backed up by more substantial data, an unusually active grapevine, publication of crucial results via the press interview rather than through established channels, a prominent part played by the popular press, and an unusual pattern of growth and decline akin to the outbreak of an epidemic. Before discussing the aspects of behavior (both on the part of those involved and those standing on the sidelines) that distinguished the polywater episode, let us trace its course from the zenith in 1970 to the eventual demise.

Polywater drains away

headline in *Nature,* March 5, 1971

8

The Noes Have It

Perhaps, from a scientific point of view, polywater should
have ceased to be a live issue once it had been established
that it was not a new and more stable form of water. However, from the number of articles published it is clear that the
controversy did not die after the Lehigh meeting. Indeed,
well into 1973 articles describing original research still appeared in the scientific journals, particularly in *Nature* and
Science, where the battle had been fought out all along.
Even when one allows for the delay in publication, it is clear
that throughout 1971 several research groups were still
actively pursuing their polywater investigations. January
witnessed a change of mind on the part of Pethica and his
colleagues: "Considering our own findings in the light of
other workers we feel strongly that anomalous water is not
polywater, and until conclusive evidence to the contrary is
obtained—the possibility still exists that it is a gel or solution
of silicates or other materials." For two years the Unilever
scientists had looked for evidence of large molecules, but
everything their analyses ever produced had a molecular
weight of 18, which is to say that it was water. They also
suggested that Lippincott's famous spectra should be regarded "with caution," because working with capillaries of
20 μm diameter required "great skill in optical alignment" of
the spectrometer. (This was indeed teaching the master spectroscopist his craft.)

Deryagin, whose reply was delayed by several months, possibly by hostile referees, took a predictably stern line. Pethica
and his colleagues were severely taken to task for showing
"incomplete understanding of and inexact quoting from the
literature." They were charged with making unwarranted assumptions and with attributing to Deryagin statements which

he had never made. The tone and content of the note show beyond any doubt that in March 1971 Deryagin was still quite unconvinced by any of the claims made at Lehigh and elsewhere that the possibility of impurities raised grave doubts regarding the existence of real anomalous water.

There came during 1971 another change of mind, more important than that of the Unilever group as it was based on persistent hard work and soul-searching. Leland Allen and Peter Kollman stated quite explicitly that the position they now adopted was "clearly opposite to that in our original work." During the whole polywater controversy this was the only honest statement by a group of workers that their *own* new findings had forced them to change their mind. Their article in *Nature* of October 22, 1971 contains no equivocations, no ambiguities, and no question marks. Curiously, there was no mention of the ever-growing experimental evidence for impurities. Allen and Kollman claimed to have reached their new conclusion strictly on the basis of some new calculations, using the same techniques that had previously convinced them that a stable form of polywater could have a real existence. Even today, however, many of the practitioners of such quantum-mechanical studies doubt whether their computational resources are yet sufficiently reliable and refined to allow discrimination between hypothetical structures that differ only marginally in energy. (In mitigation it must be remembered that Allen and Kollman were by no means alone in claiming that quantum-mechanical calculations could prove or disprove the existence of polywater. Indeed, when they had already convinced themselves that, on theoretical grounds alone, the existence of a stable polymeric water was most unlikely, others were still busy with similar calculations and were publishing reports suggesting that perhaps polywater could exist after all. Some of these theoreticians seemed completely unaware of the analytical work of Rousseau and others who had by that time cast grave doubt on the chemical purity of polywater.)

The main development of 1971 took place on the analytical front, but a grotesque element crept into the style in which the articles were written. In some quarters attempts continued to show that the properties of polywater were due to contamination by sweat. Rousseau played a leading role here, and to judge by the printed evidence he must have spent considerable time and effort to convince himself and the rest of us of the likelihood of the sweat hypothesis. In *Science* of January 15, 1971 he reported his spectroscopic measurements on sweat samples "isolated by pressing the liquid from cotton and woolen fabric that had became saturated with male upper-body sweat after extensive physical exertion." In a footnote he thanks C. L. French and C. Krupa for sweat samples. According to *Time,* on the other hand, it was Rousseau himself who provided the sweat after having played a vigorous game of handball. This was good headline stuff.

Another hypothesis that received increasing attention was that during the condensation of water vapor in the capillaries some reaction with the solid led to contamination of the water by the material of the capillary (specifically, silica). This had been suggested earlier, but several attempts to identify silicon in polywater had failed. Even now the evidence was as ambiguous as everything else connected with polywater. A research group working at the Inland Waters Branch of the Canadian Department of Energy, Mines and Resources stated that their sample of polywater contained 10 percent to 30 percent silicon. This was confirmed by Barbara Howell of the Cloud Physics Research Laboratory at the University of Missouri, who had been in on polywater research from its very beginning. Others also subscribed to this thesis— among them Willard Bascom of the U.S. Naval Research Laboratory, who also had taken this stand right from the start and had in fact committed himself to this view in a report to ONR, dated September 5, 1969, when the doubters were still very much in a minority. Bascom admits, however, that his research was "a totally biased effort to demonstrate the presence of silica." In a letter he makes a telling confession, which is applicable to much of the polywater research: "I was a seeker of truth—as I saw it."

Then there was the antisilica lobby—those who claimed, or at least implied, that the impurities in polywater originated from careless experimentation. The research group at Birkbeck College, one of the first teams to become involved (in 1967) in the examination of polywater, at last published a report (in *Nature*). The headline, ''Polywater and Polypollutants,'' was probably supplied by the editor, but it reflects the general trend. By conventional standards it is a strange report, containing no details of the work performed by the four authors but several references to such work: ''It has been well established from the work of others and ourselves that . . .'' and ''As a result of our studies and with some knowledge of the experiments of other workers, we have concluded that. . . .'' We are left guessing what these authors' own studies might have been; however, they thank nine other scientists (among them John Bernal) for help. Careful reading of the many polywater reviews of the period shows that the work of the Birkbeck group is referred to obliquely in one such review: ''Water and Polywater,'' by John Hasted, also of Birkbeck College. The trail ends here, however, because the reference is to an internal progress report, never published. Be that as it may, the Birkbeck group's reading of the available evidence—and they were certainly well aware of the situation—and ''their own studies'' convinced them that the signs of polywater originated either from gross impurities due to carelessness or ignorance or from impurity effects due to leaching from the capillaries. They claimed that if adequate precautions were taken the product found in the capillaries was indistinguishable from ordinary water, and that different anomalous properties were just reflections of different types and levels of contamination.

The accompanying *Nature* editorial ''Polywater Drains Away'' stated that ''even unhappy tales like [polywater] contain seeds of useful and constructive speculations,'' and that, on the debit side, ''The failure of several experimenters to pursue with all the vigor at their command the possibility that contamination might account for most of their observations is nothing to be proud of.''

There were also those who continued to disprove the existence of polywater long after there was still any need for this. The contribution of members of the "scientific establishment" is of interest here. I shall elaborate on this; here I confine myself to such involvement as is reflected in the published literature. This took two forms, exemplified by the attitudes of two members of the scientific elite. Joel Hildebrand, of the University of California, a physical chemist of considerable eminence, made his contribution in a two-paragraph letter to the editor of *Science*. The language was emotive:

Proponents of polywater in the pages of *Science* and elsewhere may be interested to learn why some of us find their product hard to swallow. . . . [We] are skeptical about the contents of a container whose label bears a novel name but no clear description of the contents. . . . [We] are suspicious of the nature of an allegedly pure liquid that can be prepared only by certain persons in such a strange way. . . . [We] choke on the explanation that glass can catalyze water into a more stable phase.

This was published on the eve of the Lehigh conference, and, for obvious reasons, caused a furor. At the other extreme is an article (also in *Science*) by crystallographer W. Barclay Kamb of the California Institute of Technology, who was equally eminent in scientific circles. His was an eleven-page paper, published in April 1971. One can speculate that, had it not been by an author with as much political muscle as Kamb (the son-in-law of Linus Pauling), the editor of *Science* would hardly have countenanced such a long article on a subject that was by now of only marginal importance. Despite the reference to "structural theorizing" on the part of others, Kamb's paper is really not so different in content, and can be summed up as flogging a dead horse. What were Kamb's conclusions? "Because the existence of the anomalous phase would have far-reaching implications it is important not only to test its experimental basis independently, but also to examine the theoretical explanations for consistency with structural principles and with available evidence on the forces of interactions between water molecules." One might well ask:

If polywater can be shown by experiment not to exist, then what is the point of testing its theoretical basis? Kamb's long article is supplemented with two pages containing a multitude of footnotes and well over 100 references. Its tone and content demanded a retort by Allen.

The "old" literature is sad testimony to the animosities and passions that existed, which were veiled (at times only very thinly) by current conventions of scientific publication. Allen accused Kamb, again in the pages of *Science,* of making numerous derogatary references to his first polywater paper, while ignoring the later publication, which clarified many of the questions. Jerry Donohue also saw fit to publish a reply to Kamb's article.

In general, the American, Soviet, and European scientific establishments do not come out of the polywater affair looking good. Not all of the scientists involved actually published their views in the scientific journals, but from private correspondence, remarks passed at meetings, press reports, and anecdotes one can only construct a collective attitude whose main ingredients were disbelief, mistrust, derision, and lack of grace.

Such attitudes were not confined to members of the establishment. They are to be found in the reminiscences of a few who were involved in the research, but such examples are comparatively rare, and most scientists take a much more charitable view of their fellow professionals. Of course not everyone has published his polywater memoirs, but Ellison Taylor, director of the chemistry division at the Oak Ridge National Laboratory, did see fit to do so, in the *ORNL Review* of winter 1971. He tells an amusing story of his involvement with polywater, but the flavor is one of self-righteousness: He "knew" from the beginning that polywater as such could not exist; he decided to write up his results "in an attempt to counter the deluge of misinformation and misinterpretation," and when it came to reproducing exactly the famous spectrum of polywater, as reported by Lippincott, then "everyone but us, both pro and con, was apparently more easily satisfied with just resemblance." Taylor's report of the Lehigh meeting is none too flattering to his fellow scientists. After taking swipes at Deryagin, Allen, and several others, he tells of the "off-the-cuff calculation by a Britisher [presumably Douglas Everett] of the magnitude of the surface forces which might disturb the viscosity measurements (the point I had made before, but it sounded much better with a British accent)." Predictably, this light-hearted article, although published only by an in-house journal, drew a comment from Leland Allen pointing out that a "'sense of frustration and righteousness" was not uncommon among workers on polywater. In his letter to the editor of the *ORNL Review,* Allen remarks that Ellison Taylor's attitude of presuming all along that a new phenomenon such as polywater could not possibly exist was quite inappropriate. Here then is the basic problem: The skeptics were accusing the believers of naivete and undue gullibility, and the believers were accusing the skeptics of arrogance and lack of imagination.

While all this was going on, and despite the ever-increasing likelihood that polywater would turn out to be a nondiscovery, there were still those who carried on as though nothing had happened. As late as May 1971, Barry Brummer and his associates at Tyco Laboratories sent a manuscript to the *Journal of Physical Chemistry* entitled "A High Yield Method for the Preparation of Anomalous Water." This was published later in the same year, but although the writers speculated how their process might be upscaled, events had overtaken them and there was no longer any need. Willard Bascom promptly questioned whether the stuff prepared by Brummer was polymeric water, and Brummer admitted that by now he was convinced that the liquid he had condensed by his improved process was not a polymeric form of water. Several other researchers published polywater results right into 1972.

During the whole of 1971 the debate still raged in Australia without much being added except that claims for an entirely new kind of polywater on silica-free surfaces were made. In light of the many different impurities that could apparently give rise to signs of polywater, such reports no longer excited the scientific community. Various other publications appeared that had oblique relevance to the polywater issue. As the activity gradually subsided through 1972, the reviewers and interpreters were still busy. As late as 1973 one of their number concluded in the pages of *Naturwissenschaften* (the German equivalent of *Science* and *Nature*) that the end of the polywater story was nowhere near in sight. He was wrong: On August 17, 1973, Deryagin and Churaev—in a one-column communication to *Nature*—stated that they had found it impossible to condense liquid water in capillaries that was free from impurities and at the same time exhibited the properties of anomalous water, and that therefore "these properties should be attributed to impurities rather than to the existence of polymeric water molecules." This bald statement was embellished by various explanations and qualifications, suggesting that the anomalous properties that had for so long been associated with water polymers were due to the "peculiar features of a reaction taking place between the vapor and the solid surfaces in the process of condensation." This had over the years been suggested by various scientists, but now it spelled the end of polywater.

Before the Deryagin-Churaev article appeared in print, its contents had, in the true tradition of polywater, already been leaked to the press. The *New York Times* wrote the obituary "Polywater's End" on July 28. It ends thus: ". . . recently Academician Deryagin himself has announced that his latest researches have shown the doubters were right and he was wrong. Now if only politicians behaved with the candor science requires of all true scientists."

Despite the weight of adverse evidence, and with the critical comment coming from so many quarters, it was truly surprising that the matter did not end there. In 1977 Maryna Prigogine, the wife of a Nobel laureate and a student at the Free University of Brussels, reopened the whole issue in her 200-page Ph.D. thesis. That she had not been unaware of the polywater developments is witnessed by several publications of earlier date in which she had come down in favor of the silica-impurity theory. It is surprising, then, that six years later, and without much reference to previous work, this thesis contains a *de novo* investigation of the conditions which are said to favor the formation of "anomalous" water and of its properties. Although the interaction between silica and water is recognized as being responsible for the condensation of the substance, its existence as a fluid worthy of study is nevertheless taken for granted, and Deryagin's theory of its formation is once again analyzed before being finally discarded. A whole chapter is devoted to the description of experimental work and to discussions of the most favorable conditions for, and the mechanism of, formation of "anomalous water." In its style the thesis is reminiscent of Deryagin's early publications, and, reading it, one has the distinct feeling of déjà vu.

Apart from this one significant revival, activity on polywater ended with the inevitable post-mortems in the press: one in the *Washington Post,* one in *Science,* and another one in *Chemical & Engineering News.* The latter article, which emphasized the virtues of keeping an open mind and being ready to change one's opinion on the basis of evidence, quoted a letter from a reader: "The real value of the polywater situation is that it has helped to reveal the true nature of a scientific controversy. . . . researchers know that any similar dispute shares the same acrimony, disputed results, and mistakes of good men earnestly seeking the truth." This is a positive and charitable view that is echoed in Leland Allen's assessment of the polywater affair in *New Scientist*. Allen is unrepentant as to his own role: "I was strongly stimulated by both the human and scientific experiences, and the attention focused on my first paper on the subject gave me a more convincing sense of having made a true contribution to science progress than I have felt for some other conventionally successful research projects." Allen also believes that the quick resolution of the phenomenon bears witness to the efficiency of the scientific method. This view is by no means held unanimously.

Experience is not what happens to a man. It is what a man does with what happens to him.

Aldous Huxley

9

Hindsight

While the polywater debate was raging—when any moment a telephone call might bring news of the latest results—few of the researchers could take time to reflect on their own involvement in the study of polywater. And did not the fact that they could drop whatever they had been working on mean that it was unimportant in comparison with polywater? The situation is different now. Not only have we witnessed the fate of polywater, but there has also been a profound change in the ground rules that govern the funding and the conduct of scientific research. In these days of project and mission orientation and of strict accountability, it would be much harder to switch resources to pursue, at a moment's notice, the sort of problems that were posed by Deryagin's anomalous water.

Nevertheless, polywater is not totally dead, since several of the questions it raised have not yet received satisfactory answers. Some of these questions have, on the face of it, considerable scientific merit. For instance, the indications were that water *vapor* reacts with quartz more readily than does liquid water. Philip Low of Purdue University suggests that "in their haste to dissociate themselves from anomalous water, members of the scientific community are ignoring certain important problems." For example, if (as is now accepted) the properties of polywater are due to high concentrations of siliceous material, how could such large quantities of silica be dissolved from quartz and Pyrex capillaries? Quartz is certainly not sufficiently soluble in water to yield the observed silica concentrations, a point made repeatedly by Deryagin in his rebuttals of charges that silica was responsible for the observed properties of polywater. This raises valid questions: Is water adsorbed from the vapor phase onto silicate surfaces a much better solvent than bulk water? Is it more acidic than bulk water? This has certainly been suggested from time to time. More generally, the vexing question of the modification (albeit not permanent) of water in the proximity of surfaces is still very much with us. That such effects exist is now beyond doubt, but their adequate analysis in molecular terms has not yet been possible. There are also related and unsolved problems in defining the role played by water in the maintenance of the so-called native states of biological macromolecules. The questions associated with "bound water" are also still very much with us. The solution of all these problems has probably been delayed as a direct result of what happened during the polywater period. To the initiated it has long been known that experiments in surface science demand much time and effort, exactly because of the disproportionately large influence of impurities on the measurements. It does not therefore require much in the way of discouragement to stop all such investigations.

Any researcher still courageous enough to invoke the existence of "modified" water must be prepared to face a good deal of ridicule. The funding agencies, too, do not look kindly on any proposal that smacks of polywater. Prospects for such work are not improved by those researchers who, in their speculations, venture beyond currently acceptable bounds. The polywater experience has had a sobering effect on all those who had been and still are active in water research. We have all been sensitized by our contact—be it ever so tenuous—with polywater. This has made us more conservative and more reluctant to listen to suggestions of different forms of water. We automatically dissociate ourselves from anything that might bear a resemblance to polywater. These reflexes are perfectly understandable in view of past history. Unfortunately, polywater has left its shadow on a large area of physical chemistry and biochemistry concerned with solvent effects. This becomes obvious during most discussions on such topics: Sooner or later a jocular reference will be made to polywater.

The majority of those who first became infected by polywater had been active in the general area of surface chemistry. This was natural, because they would have been the first to appreciate the significance of Deryagin's hypothesis, and they would also have been familiar with the experimental techniques employed in the preparation and study of anomalous water. It is gratifying that most of those concerned have not given up their interest in surfaces (a fascinating and important branch of science) and have been able to continue their work. Such scientists are in a good position to look back on their own involvement with polywater as a brief interlude in their long-term research efforts. Many of them have been generous in sharing their experiences with me and allowing me to quote from their letters. Before going into detail, let me summarize what seem to be common ingredients in most of the reminiscences.

Scientifically, the 1960s were a period of great excitement, particularly in the United States. I was privileged to be able to spend a year at one of the centers of scientific activity during that period (before the days of polywater), and was very much influenced by this stimulating atmosphere. One felt that anything was possible. The scientist had an enviable position in society; he was respected and admired, he was afforded generous facilities, money seemed to be inexhaustible, and the scrutiny of grant proposals was fairly superficial. Scientists were free to travel extensively and to meet and entertain one another at congresses, often in exotic surroundings. The stringent financial controls and the intense competition for funds and tenure that has produced the ''publish or perish'' mentality were still problems of the future. There were of course a few prophets who tried to convince their fellow scientists that the end of their millennium was at hand. In a lecture Donald F. Hornig, then scientific adviser to the president of the United States, touched on the subject of accountability. He suggested that it might well be in order for professional societies to hold their annual congresses in faraway places like Rio de Janeiro. All that had to be done was to explain to the taxpayer the benefits to society (or, indeed, to scientific progress) of holding such a meeting in Brazil rather than in Atlantic City; no doubt the taxpayer would understand. In 1963 Hornig was one of the minority of scientists who knew or sensed that things might change.

The soil was well prepared for what was to happen. Only a handful of American scientists were influenced directly by the work of Deryagin. In the 1960s the Iron Curtain still extended well into the realm of science, and the Russian scientific literature was largely unknown and ignored. Most Americans were alerted to Deryagin's discovery by Ellis Lippincott's reports of his experiments in *Science.* If one single article can be said to have been responsible for all the subsequent developments, it is the one by Lippincott, Stromberg, and their partners. Similarly, if we look for the origin of the extraordinary interest of the media in the affair, we find that it stems from the warning by Donahoe that polywater was probably the most dangerous substance on earth.

Most of my correspondents display a distinct nostalgia in their reminiscences. The recurring general theme is: "It turned out to be an artifact, but what if it had been true?" It would have been just one more example of science fiction having been turned into science fact twenty years later. Thus, David Goring of the Pulp and Paper Research Institute of Canada may have played only a minor role in the polywater affair, but is not at all ashamed of it. In fact he wishes that we should have more polywater episodes, because "science today seems to be dominated by the conventional wisdom of the great or once great. We lack entrepreneurs in science. In proving them wrong we learned a lot. Sometimes they turn out to be right."

Another common feature of the recollections concerns the sense of urgency with which people set about repeating Deryagin's experiments. The methods looked so simple and did not require expensive equipment. Milton R. Lauver of NASA believes that this was one of the most attractive features of polywater. In contrast with space research, which was then flourishing and which required the battalion approach, one- or two-man teams could make significant contributions. In fact, any undergraduate should have been able to prepare polywater. Indeed, several university departments arranged lectures for their undergraduates during which details of the experimental techniques were provided for aspiring do-it-yourself polywater researchers. Then again, most of my correspondents admit to somewhat sketchy experimentation, and several had trouble getting their results published. This was important: The stir created by Lippincott's publication had shown clearly that there was a premium on being first in print.

Most scientists will admit readily that the most important motive for doing research is to establish priority (hence the accent on so-called *original* research), or at least to present evidence of independent discovery. This is referred to as the Eureka syndrome. Historically, the emotional and intellectual driving force is said to be the elation of discovery, as documented in the writings of some of the greatest scientists of the past. In more recent times, as sociologists have suggested, the pure gold of "elation of discovery" is more often than not alloyed with the baser substances of egotism and of craving for recognition by a peer group or by society at large. Was this the reason for the many polywater press releases? In the realm of science, discovery *is* property; once the discovery is made public, the knowledge becomes communal.

Quick and perhaps premature publication was one of the characteristics of the polywater work, and it is a theme that recurs frequently in my correspondents' assessments. But they differ in the motives which they ascribe to this practice. Pieter Hoekstra takes the charitable view that the driving force was the desire to enter the mainstream of excitement— in other words, the Eureka syndrome. Philip Low, on the other hand, discerns the propensity of scientists to "enhance their prestige by being identified with a new and exciting discovery that is winning popular support." John L. Anderson of Carnegie-Mellon University suggests that the journal review system broke down, since many of the "me too" papers which followed Deryagin's original publication added little or nothing positive to the controversy.

Since research is supported almost exclusively by public funding, it seemed of relevance to this discussion to establish the approximate amount of money that was devoted or diverted to polywater research. The attempts were singularly unsuccessful. Apart from acknowledging the grants allocated by government agencies, few will admit to having spent identifiable sums on polywater research. Words like ''bootlegging'' and ''moonlighting'' occur in answer to my question about funds devoted to such activities. This again indicates that the individual efforts were of a minor nature and that the Western countries never mounted an organized effort in any way comparable to that of Deryagin's group in Moscow. Perhaps the funding agencies regarded the support of polywater as a kind of insurance premium. During the hey-day of polywater I had occasion to ask the then-secretary-general of the U.S. Office of Saline Waters why he had seen fit to allocate approximately 10% of his annual research budget to support polywater activities at a number of institu-tions. Did he believe that polywater was a genuine discovery, that it had a future, that there were possibilities of exploita-tion? He replied that he had not been convinced of any of this, but that if it had turned out that polywater was ''real'' and that his agency had not supported research, then. . . . (He did not finish the sentence, but in a self-explanatory ges-ture drew his finger across the throat.) Most of the funds seem to have been allocated on a very *ad hoc* basis, with no commitment for a longer-term program. This made it easy to discontinue support as soon as the balance of the evidence turned against polywater.

The responses to my questions illustrate many of the facets of a scientist's life in a research institution. They describe the stimulating atmosphere of the late 1960s, the maneuverings and personality clashes, the excitement of a discovery, and the frustrations of rejection by referees. Willard Bascom, of the Naval Research Laboratory, has been particularly forthcoming in detailing his involvement with polywater. Apparently the NRL first got wind of the Russian developments in 1966–1967, and realized the military implications immediately. Bascom and his colleagues responded to the enquiries from Navy Intelligence that anomalous water was probably the result of a reaction between water and some silicate product in the capillary; nothing more was heard until mid-1968, by which time Ralph Burton was sending his reports from London. James Schulman, then Director of Materials Science at the NRL, had a long-standing interest in the structure of water and was therefore receptive to suggestions that anomalous water was worth closer study. In addition there was the possibility of attracting financial support for such work from ONR, and this consideration was, according to Bascom, "no small part of the driving force to undertake research on polywater at NRL." Opposition to this course of action came from two other eminent scientists, William A. Zisman and the late Curtis R. Singleterry, both of whom had by that time retired from management at the NRL but were still active as consultants. They had over the years engaged in a long-standing dispute with Deryagin about the existence of long-range forces between solid surfaces. Bascom writes, "I wish that I could say I had an unbiased opinion but I had spent my entire professional career at NRL with Singleterry as mentor. We viewed Deryagin as an exceptionally gifted experimentalist who frequently misinterpreted his results. . . . At the time the polywater caper broke both Zisman and Singleterry . . . were unable to do more than advise that work on polywater was a waste of time and it would be shown to be an artifact soon enough." Nevertheless, James Schulman persuaded the then head of the NRL Chemistry Division, Ronald Kagarise, to release Bascom from his research on glass fiber surfaces in order to investigate polywater. This presented a slight management problem, because Bascom was at the time working toward his Ph.D. degree and some of the glass-surface studies included experiments which would constitute part of the thesis. As is usual in these affairs, a compromise was struck that satisfied the two managers and caused dismay to the subject. It was decided that Bascom would continue his thesis simultaneously with the polywater work. Being then nearly 40 years old, Bascom was hardly pleased by this course of events. Also, he recalls that "it did not help having Bill Zisman come by my office on occasion, clucking about what a waste of time it was to work on polywater."

The plan was for Bascom to prepare large quantities of polywater, while Schulman would prospect for funds to set up a major program with a view to characterizing the material fully. Bascom emphasizes that Schulman and others at the NRL were by no means ignoring or discounting the views of Zisman, but the pressures were such that they had to balance Zisman's advice against Deryagin's experimental evidence, the opinions of many competent scientists that such an anomalous form of water might indeed exist, and the sheer scientific excitement of the concept. When news of Lippincott's spectroscopic studies reached NRL via the grapevine, this really tipped the balance in favor of polywater.

Bascom started, like most of his contemporaries, by exposing capillaries to water vapor over saturated salt solutions and waiting for columns of the anomalous liquid to grow. He soon realized the dangers of contamination and refined his techniques. By autumn 1969 he reported to ONR that all of the properties of the liquid, other than Lippincott's spectrum, could be explained on the basis of the silicate contamination. At this stage he also gave a telephone interview to *Industrial Research* suggesting that polywater might well be a silica sol. "From that time on," he says, "I was swamped with phone calls and letters requesting further information and/or offering unsolicited opinions. I understand that everyone else working on polywater had the same experience."

In 1969 the doubters were in a minority. Bascom was given to understand by professional friends and colleagues that his attitude "was born of stubbornness and would be a professional disaster when the case for polywater was proven." He candidly admits to a total bias in his work to demonstrate the presence of silica and to find an explanation for Lippincott's spectrum in terms of silica contamination. He was therefore pleased to see in *Nature* a report to this effect from a "competitor," Arthur Cherkin. This was the first indication of the presence of silica in polywater. Bascom wrote to Cherkin expressing his own agreement and the hope that the report might "serve as a brake on the runaway speculation about polywater." This hope proved to be too optimistic. In his reply Cherkin took a more realistic line: "I doubt that anything less than a convincing experimental demonstration that polywater contains silicon will do much to deter speculation. I therefore am hopeful that your own experiments will resolve the leaching question, one way or the other."

As far as Bascom was concerned, the main objective was now to find an explanation for Lippincott's famous polywater spectrum. Singleterry suggested various ways in which contaminants could affect the infrared spectrum of water, and eventually Bascom was able to "reproduce" Lippincott's spectrum by the judicious addition of extraneous substances to water. However, these results, published in an internal NRL report, did not receive more general publicity until 1972.

Bascom recalls a visit to NRL by Deryagin. After giving a lecture saying little about anomalous water, but stressing the existence of long-range surface forces, Deryagin met Zisman. Since Deryagin was a foreign citizen, he had at all times to be accompanied by a designated escort. Willard Bascom was given this assignment, and so had the unique experience of witnessing the encounter between two acknowledged giants in the world of surface science. He recalls: "Dr. Zisman reviewed the accomplishments, both fundamental and applied, of the surface chemistry group at NRL. Professor Deryagin nodded politely at the appropriate times. Eventually Dr. Zisman suggested that we had serious doubts about polywater. Deryagin responded that he had been investigating it over the past nine years and currently had twenty-five scientists working on it and he would not put that much effort in it if it were an artifact. The conversation ended on that note." A later visit with Singleterry ended equally inconclusively, with Deryagin suggesting that Bascom had not been sufficiently careful and had contaminated his anomalous water.

It is hardly surprising that these encounters did not lead to any conversions; they rarely do. All that can be hoped for is that an exchange of views might suggest to both parties some new approaches, or some definite experiments. Sometimes such discussions can lead to an agreement for collaborative studies, but the polywater episode was curiously lacking in collaboration between different groups. Everyone was playing his cards close to the chest.

Barbara F. Howell, now at the National Bureau of Standards, became involved in the polywater controversy early in 1969 through a postdoctoral appointment with the Cloud Physics Group at the University of Missouri, which had just been awarded a grant from the ONR to study the preparation and properties of polywater. She was then 45 years of age and the mother of three teenage children, two of them in college. In order to take up the fellowship, she had to part temporarily from her husband and move with her daughter to Rolla, Missouri. Like everyone else, she began by preparing anomalous water, using the conventional methods. Interesting effects were observed, but Howell was not satisfied with the purity of the substances she observed in her capillaries. With the help of an "enormously patient undergraduate student" she succeeded in milking a large number of capillaries to obtain a good-sized drop. Unfortunately, this turned out to be ordinary water. Once in a while strange spectra were recorded, but nothing could ever be reproduced, and Barbara Howell suffered from the same sense of frustration that was the lot of most polywater experimentalists. After ten months' work she was still uncertain about the existence of the elusive fluid and therefore decided to build a more elaborate apparatus for its production. This also turned to be a failure; the apparatus leaked and no anomalous water was ever generated. Like most of her colleagues, Howell attended the Lehigh symposium and came away bemused by what she heard. Once again she set to work and constructed a refined piece of equipment to generate polywater. This time she was successful, but from the properties of the fluid she had managed to collect she had to conclude that it was not anomalous and that it contained a range of impurities, in particular silicon.

By now the fellowship was coming to an end, and Howell submitted her results for publication. Her experiences were typical: "The manuscript was severely criticized by the referees and returned." She later found out that one of the referees was Russian and the other one, although American, had an important stake in the existence of polywater. After much rewriting and further disputes with referees, the editor finally decided to ignore their views and to publish the work, but by that time (thirteen months later) it was no longer news.

Despite the frustrating experiences, Barbara Howell recalls her stay at Missouri as "scientifically the most interesting chapter" in her life. Looking back over the two-year period, she is surprised that so many people could have overlooked some very simple thermodynamic facts in their search for polywater.

Robert Good of the State University of New York at Buffalo, an experienced surface chemist, was also unable to prepare polywater under conditions in which extreme care had been taken in the cleaning of the apparatus. He reflects that open-mindedness (including his own at the time) was "carried too far, but not much too far." He criticizes the theoreticians who by their conduct "proved that they can explain anything if they want to badly enough, including why a nonexisting substance is stable." On the broader issue he thinks that "we— the scientific community—do not understand scientific methodology. There is no *explicit* account available, other than the obsolete and/or incorrect accounts written by the philosophers.'' Good rejects current theories of contemporary empiricism and positivism, which he considers ''at best useless to scientists.'' He also claims that Thomas Kuhn's ideas about how scientific progress is made, as outlined in his classic *The Structure of Scientific Revolutions,* are little better in helping scientists to understand why they act in the way they do. He wonders, if scientists do not understand their own motives, how they can know how they *should* conduct themselves. Here then is an example of the gulf of understanding between the practicing scientist and the philosophers of science, and this is far from being an isolated example in my correspondence.

A. C. Hall of the Mobil Research and Development Laboratories at Dallas, Texas believes that the interested scientists were split three ways over the issue of polywater. There were those who were convinced that it was a completely new kind of water, and that it therefore merited almost unlimited investigation. Then there was the large majority who had misgivings but felt that the scientific pursuit of polywater might yet produce valuable knowledge. Only a few saw anomalous water as an aberration. However, scientific simple-mindedness could hardly be cited as the reason why widespread belief in the existence of polywater continued for so long. Hall believes that the reasons for this were twofold. It all happened during a time when ''pressure to publish and competition for funds had already gone far in creating within the scientific community the kind of breakthrough mentality which was all too ready to welcome and exploit almost anything that had potential publicity value.'' It seemed to Hall not a coincidence that some of the major polywater projects were in so-called high-technology institutions which, depending almost entirely on public funds, were quite prepared to advocate the absolute necessity of keeping up with the Russians. The other factor identified by Hall for the initial gullibility of scientists was that the advent of polywater followed closely the discovery in 1962 of chemical activity among the rare gases: "The collapse of a universally accepted hypothesis must surely have inclined many people to lose not only blind faith in other solid traditional assumptions but also, unfortunately, healthy skepticism regarding shaky new ones." It is hard to judge whether the recognition that the rare gases were not inert had any direct link with the attitudes displayed during the polywater debate, but it does serve as an example of how accepted ideas can be overturned on the basis of incontrovertible evidence. However, such evidence was lacking in the case of polywater.

Hall, like many others, had great difficulties in getting his experimental results accepted by the scientific journals. At the demand of a referee he had to moderate his "rejectionist" interpretation before his work was accepted. Stephen L. Kurtin of the California Institute of Technology had similar experiences. He became interested in polywater as the result of his wife's chance remark at a dinner party, and subsequently was responsible for some of the analytical attempts to establish the electronic properties of the substance. He met the characteristic "yours may be contaminated but mine isn't" attitude, coupled with reluctance by *Science* to publish his results. In an afterthought he quotes Nobel laureate Richard Feynman: "There is no such thing as polywater because if there were, there would also be an animal which didn't need to eat food. It would just drink water and excrete polywater" [and could use the energy difference to maintain its metabolism].

For Pieter Hoekstra, who had for several years been studying the adsorption of water on clay minerals, the involvement in polywater seemed quite natural. He also sees the developments as a reaction to the privileged position enjoyed by space scientists during the 1960s: "Here was a topic in which everyone could join in. To participate all you needed was glass capillaries, a microscope, a camera, and perhaps a refrigerator. For once scientists were able to leave their pigeon holes and cross their specialized disciplinary boundaries." Hoekstra thinks that another reason for the immediate popularity of polywater was that the scientific issues seemed simple and clear-cut. Even local newspapers ''could expound on the subject with great fervor. Their source of information could be the local college professor.''

Some of the more exotic flights of the imagination for which polywater was responsible are recalled by Walter Madigosky of the Naval Surface Weapons Center. In common with most others, he first heard about anomalous water from Ralph Burton's ONR-London reports. Later, Lippincott and some of his students visited his laboratory. Although they gave little away (no spectra were shown), they left the message that polywater was real and that it was not water. Like everyone else, Madigosky and a colleague began by drawing capillary tubes and suspending them in desiccators exposed to water vapor. Soon Madigosky realized that there must be several institutions doing just the same thing in the Washington area, because Fisher Scientific had run out of desiccators. His work was ''frustrating in that our efforts produced very little yield.'' Whatever spectra were obtained could never be repeated. Eventually Madigosky found that sodium acetate reproduced the Lippincott spectrum. By this time the ''polywater versus sweat'' debate was in full swing, but Madigosky was allowed to publish his results in *Science* after shortening his paper severely.

It was around this time that the first men landed on the moon. The astronauts reported moon dust sticking to their boots and clothes. Madigosky recalls a television interview at the time in which two young scientists speculated on the existence of polywater on the moon. They argued that its properties were quite consistent with the astronauts' reports. This is an example of the atmosphere in which the polywater work was performed and reported.

The twin problems of self-deception and wishful interpretation of results are touched on by Leland Allen, who graphically describes the dilemma which he faced in 1969. On the

one hand were the results of Deryagin, now backed up by
the spectrum recorded by Lippincott. On the other hand, he
knew of the analytical studies then being performed by Rousseau at Bell Laboratories. If the charge of careless experimentation could be leveled at Deryagin and Lippincott, Allen felt
that the same was true for the experiments of Rousseau. At
the very least, they suffered from that hastiness in concep-
tion and prosecution so common in most polywater research.
Perhaps even more important were the general climates in
the two laboratories concerned. At the University of Maryland there was a firm belief in the existence of a new form of
water, whereas Rousseau's colleagues and superiors at the
Bell Laboratories were just as firmly convinced that no such
substance existed. Knowing this, Allen felt that "one was
forced to discount most of the claims on both sides." However, on balance he thought that polywater was a true "and
even typical" scientific discovery; hence he decided to investigate its possible structure, with the aid of theoretical techniques which had been reasonably successful in other cases.
Allen therefore steered a middle course between those who
were convinced that polywater was a genuine discovery and
those who, right from the beginning, had been committed to
demonstrating its artifactual nature. This being the case, Allen justifiably claims that the issue was handled reasonably
well by the scientific community. He is not alone in holding
this view. Milton Lauver also observes that "the international
scientific community acted in a proper and responsible manner in reporting the research results in the usual journals
without editorial comment," and that "knowledgeable individuals reached their own opinions on the basis of the facts
disclosed."

The opposite view is held by several other scientists. Sefton
Hamann of Melbourne, Australia tried hard but without success to convince his fellow Australians that their "new" polywater, produced in the absence of quartz, was probably nothing but magnesium carbonate or magnesium hydroxide. He

argued strenuously against unquestioning acceptance of the infrared spectra, also without success. It is not surprising, therefore, that Hamann views the episode as ''the darkest and most miserable in the history of physical chemistry.'' Nor is he alone in holding such views. John L. Anderson of Carnegie-Mellon University asserts that the scientific community should have required much more positive proof of the existence of polywater before granting it the status it did. There was altogether too much negative proof, a bad sign. Anderson is still dismayed by the number of scientists who were ''taken in by the publicity campaign waged by Deryagin and others.''

Perhaps at that time scientists *wanted* to believe and to be part of this amazing discovery. How else can one account for the almost complete disregard of impurity problems, which, apparently, were only realized fairly late? The seriousness of contamination is one of the most important problems facing surface scientists. John Finney, in a 1973 obituary of polywater, marvels at the lack of communication about the issue. It was almost as if scientists did not want to know about the impurities. Finney concludes that the scientific community had the know-how for the proper application of the scientific method. A half-dozen research groups could have resolved the problem surrounding anomalous water. Instead, upwards of 400 scientists became involved. He asks ''Is this efficient?'' and claims that ''we should have done much better.'' Coming from one who actively participated in the polywater affair, such criticism is perhaps harsh, but it is fair. Indeed, we should have done better, but not because the issue was handled in a wasteful manner. Efficiency is not one of the characteristics of scientific research, and polywater was not really outstanding in the way resources were misapplied. In terms of money spent—if this is the yardstick for efficiency—polywater probably produced more in the way of results and publications than most scientific activities. That the results turned out to be misleading or wrong and that many of the publications were of little intrinsic value is a different matter. It is this that should give cause for concern.

Anybody who believes that the allocation of resources for scientific research is based on a policy conceived in a rational manner should read J. J. Salomon's analysis "Science Policy and its Myths" in the journal *Public Policy.* A number of the issues discussed by Salomon apply to the handling of polywater (although in a negative sense, because research activity associated with polywater was never planned; it developed like a gold rush, and ended in a similar manner). Salomon holds that peer judgment in science, both for funding and refereeing, is the nearest thing to a market economy, that is, a system based on supply and demand. Its disadvantages are inertia, vested interests, dogmatism (by the peers), and discrimination against young people and new ideas. None of these claims could ever be tested in the polywater affair, because it did not last long enough for any policies to be developed. On the one hand polywater was considered "useful"; hence the great interest taken by ONR in the initial stages and the funding irrespective of any peer judgment. On the other hand, laboratory activity did not require large sums and could be pursued irrespective of the opinions of the peers. It is impossible, therefore, to assess efficiency and effectiveness in the polywater affair. Presumably, activity in Deryagin's laboratory was carefully planned; certainly the research output was prodigious. Were the Russians therefore more efficient?

Even among those who were actively engaged in polywater research, there is no agreement about the role of Deryagin, the nature of the response by the Western scientific community, the quality of the experimental and theoretical work, the motives of those who got involved, the role of the scientific journals in publishing or in refusing to publish so many of the minor contributions, or the tone of the scientific dialog. On two points there is general agreement: The early involvement of the popular press was a pernicious influence that made it harder for the issue to be resolved by the normally accepted methods, and the tone adopted by several of the protagonists and by many of those on the side lines was unhelpful. Philip Low states that ''it is the obligation of a scientist to criticize unreliable data or interpretations, but his criticism should be reported in an objective, scholarly and gentlemanly way. Otherwise it should not be published.'' While such high standards may be difficult to live up to, with polywater we experienced unacceptable deviations. In Low's judgment, ''the condescending, disdainful, demeaning and sarcastic remarks about anomalous water, its originators and supporters which appeared in the literature, especially after it became suspect, were unbecoming of scientists and should never have been published.'' These are harsh words, but if scientists want to maintain that the pursuit of knowledge and truth is their stock in trade, then they must be seen to be above petty personality squabbles in their written discourse.

Ten years after the events, the consensus among the scientists involved is certainly not one of unmitigated shame. As might be expected, there is a certain degree of ambivalence, but on the whole we appear to have emerged from the affair without too much discredit, and certainly somewhat wiser for the experience, which has thrown into sharp focus the pressures—external and internal—that mold scientists' work. Without doubt the polywater episode provided the general public with a glimpse of the world of the scientist and revealed something of the ''human'' side of scientific research. Perhaps more important, it also provided a mirror in which the scientist himself could see reflected and caricatured some of the aspects of his profession which are usually taken for granted.

The assessment of the polywater affair is not complete without some attempt to place it in a more general sociological context. Indeed, it could be argued that the scientists who were deeply and personally involved in the affair (whether for or against polywater) are, even ten years later, not the best judges of its significance. Eric Larrabee took our profession severely to task when he wrote that ''scientists seem to be able to go about their business in a state of indifference to, if not ignorance of, anything but the going, currently accept-able doctrine of their several disciplines . . . . The only thing wrong with scientists is that they don't understand sci-ence. They don't know where their own institutions came from, what forces shaped and are still shaping them, and they are wedded to an antihistorical way of thinking which threatens to deter them from ever finding out.''

Fair is foul, and foul is fair.

Shakespeare, *Macbeth*

10

A Case of Pathological Science?

Even before the fate of polywater was sealed, there were accusations and recriminations both in print and in private. They originated both from laboratories that had become centers of polywater research and from individuals who had stayed aloof but had never taken great pains to disguise their distaste for the whole affair. Many of these people now came out into the open with "I told you so" commentaries. The immediate reaction was that some of the researchers who only a short time ago had been hard at work trying to prepare, isolate, and investigate anomalous water now protested that they had been motivated purely by the scientific challenge. (Of course, they were not always charitable enough to judge the motives of their fellow scientists similarly.) Words like "deviant," "pathological," and "shoddy" were used. Some of the harshest criticism came from members of the scientific elite; taking their cue from such establishment figures, the editorial writers of the popular scientific journals were quick to pass judgment.

The newspapers soon dropped polywater (it had ceased to be news), but the scientists have not been able to forget it. Even now the issues raised by polywater are still lively topics for discussion over coffee in the laboratory, and the topic invariably crops up during question time after seminars dealing with more mundane aspects of water. As both a scientific and a sociological phenomenon polywater made a major impact on practicing scientists at large. Since the label "pathological" has been used on several occasions in connection with polywater, we need to enquire into its validity in this context. We must first establish the symptoms of pathological science and then examine whether and how polywater differed from earlier, similar episodes in the history of science.

When polywater is referred to as ''pathological science,'' what is really implied is that the conduct of scientists showed some abnormal behavior patterns. It is claimed that these became apparent in the manner in which scientists became motivated to engage in polywater research, in the way such research was performed, in the methods of communication chosen, and in the scientific dialog that developed. Without doubt, the involvement of those who were themselves not part of the polywater community, including the news media, further complicates such an analysis.

I am well aware that I am treading on dangerous ground, because we now cross the frontier that separates the territory of the professional scientist from that of the sociologist of science. Anything I might be able to contribute to a better appreciation of the significance (if any) of the polywater episode may be trivial and amateurish in comparison with the learned analyses of similar events that can be found in the professional literature of the philosophy and sociology of science. In fact, it may all have been said before, although I doubt that. Certainly, if polywater is in any way different from past episodes (and I believe that to be the case), then a definitive verdict by the sociologists must be left for the future. It is much too early for such judgment to be made, because most of the *dramatis personae* are still active in the profession and the dust has not yet been allowed to settle sufficiently. However, since the final judgment on the significance of polywater as a socioscientific phenomenon will eventually be made not by the scientists but by sociologists of science, it is necessary at this point to examine briefly the relationship between practitioners and sociologists.

It seems a fact of history that, whenever a profession has established itself, and when its practitioners have created for themselves a framework of theory and practice, there develops in their midst a group of individuals whose main interest becomes the study of the practitioners at work. The emphasis is on analysis and interpretation, and on placing the activities of the practitioners in the context of society as a whole. Initially, as such a group of interpreters and analysts develops, all its members are drawn from among the practitioners. Many of the early philosophers of science were famous scientists and mathematicians; they were in fact "natural philosophers." They spoke the language of their fellow practitioners and wielded considerable influence on their work and its developments. However, if the analysts and interpreters are at all successful, then sooner or later they establish their own departments within universities, where students can then study the philosophical and sociological aspects without having ever been part of the community of scientific practitioners. In due course the vocabularies of the two communities diverge, and after a few "student generations" the links between those who practice and those who analyze become tenuous.

This is not to gainsay the value of developments in the sociology and philosophy of science in helping us to understand fundamental principles and the place of science in society. In principle, such an understanding could lead to a profound influence on the part of the sociologists on the future development of the profession they study. This has certainly been the case in the field of education, where we have witnessed the growth of a veritable industry, many of whose members, though they have never faced a class of children, are involved in such matters as school organization, pupil selection, curricula, and examination procedures and exert considerable political influence. The influence of educationalists on education is in no way matched by that of sociologists on the development and organization of scientific research. Indeed, what sociological developments there have been have hardly influenced the thinking of scientists or their training as professionals. If the sociologists of science wielded as much power as their colleagues in education, then scientific research today would be conducted quite differently, although not necessarily more effectively or to better purpose.

The divergent viewpoints taken by practitioners and sociologists of science derive from their respective roles as "insiders'' and ''outsiders,'' a classification first defined by Hegel and more recently discussed in the context of science by Robert K. Merton. Both groups contribute to knowledge, but in very different ways: one through first-hand, personal experience and the other through academic means. The English language does not permit this distinction; the word *knowledge* describes both aspects. In German *(kennen-wissen)* or French *(connaître-savoir)* this distinction between insider and outsider is quite clear. In the realm of science the practitioners are the insiders and the philosophers and sociologists the outsiders. This is only a first-order approximation, but it will serve. The curious fact is that few of the rapidly growing number of insiders seem to be motivated to acquire the necessary academic background knowledge to be able to judge dispassionately the historical development of science, its institutions, and its role in present-day society. It might of course also be argued that the very exposure to active science makes the insider unfit as a dispassionate judge of events. This charge has on several occasions been leveled against Leland Allen, even though the printed evidence shows clearly that he at least tried to grapple objectively with the professional, ethical, and scientific issues raised by the polywater affair. Nevertheless, while the dispute raged, he was one of the most committed insiders, and one might question how completely one can shake off past personal experience and emotional involvement.

Regrettable though it is, there is now a gulf separating the communities of scientists and sociologists of science. The student of philosophy and sociology who has never been exposed to life in the laboratory cannot begin to appreciate the subtleties and rituals governing the conduct of research, the intercourse between scientists, and the hopes, desires, fears, and ambitions that form part of their everyday lives, insofar as they differ from those of other groups that make up society. This lack of personal involvement and appreciation makes some of the sociologists' writings appear academic and dry, full of statistics and with little human interest. We can learn how many papers are published annually in a journal, the age distribution of the authors and referees, the percentage rejection of papers in different disciplines, or the number of citations a given paper has attracted, but the mountains of quantitative data are hard to digest and to translate into qualitative evaluation of the originators of the data.

On the other side of the gulf, the student of science and technology goes about his business without much notion of the historical roots of his discipline. Mention has already been made of Eric Larrabee's impatience with the antihistorical approach of scientists toward their discipline. There is much truth in his comments; most scientists confine their reading to a very narrow, specialized literature, written in the jargon of their subdiscipline and of the day. Leland Allen, who has devoted much thought and soul-searching to polywater and its implications, agrees with Larrabee when he writes that "as scientists we do not yet have the perspective of sufficient time to get over the prejudices and fashionabilities that influence our everyday behavior. We also tend to discount or oversimplify historical, sociological, and psychological aspects. We are too operational and too win-lose oriented. We come out simplistic and opinionated."

It is likely, then, that the majority of scientists find it hard to understand the language of the sociologists of science. Even if they do understand it, they may well disagree with the outsider's sociological analysis of an issue such as polywater. Yet, at the same time, there is much substance in the point that the practicing scientist knows little and cares even less for the history of his discipline and for the forces that shape his profession. It is not a foregone conclusion, however, that a practicing scientist's view of the polywater affair must of necessity be more biased or less valid than those of the philosophers and sociologists. Indeed, my justification for discussing such matters at all is that, with "insider" experience in regard to the conduct and management of research and the peculiarities of water, I may look at some of the questions and issues rather differently from those whose main concerns are the purely philosophical and sociological aspects.

Turning now to the aspects of polywater that might be labeled as pathological or at least unusual science, I must state at the outset that it is impossible to divorce the properties claimed for the substance from the quality of the experimental observations on which such claims were based. In the early days, before the situation became confused by various nonscientific factors, anomalous water was just a laboratory finding that had to be taken at face value, tested, evaluated, and explained. Even then, there were already various well-documented case histories of scientific discoveries that had started in a very similar manner but had not stood up to such examination. Indeed, in 1953, many years before the first reference to anomalous water appeared in the scientific literature, Irving Langmuir defined pathological science as ''the science of things that aren't so.'' He analyzed several historical examples and suggested six criteria:

- The maximum effect that is observed is produced by a causative agent of barely detectable intensity, and the magnitude of the effect is substantially independent of the intensity of the cause.
- The effect is of a magnitude that remains close to the limit of detectability, or many measurements are necessary because of the very low statistical significance of the results.
- Great accuracy is claimed.
- A fantastic theory, contrary to experience, is put forth.
- Criticisms are met by *ad hoc* excuses.
- The ratio of supporters to critics rises to somewhere near 50% and then falls gradually to oblivion.

Going back through the scientific literature, one comes across several discoveries that by these criteria can be classified as pathological science. A well-documented and fairly recent example is the case of the so-called N rays, which were a by-product of the intensive research that followed the discovery of x rays. In 1903, Blondlot, a member of the French Academy of Sciences, described a new type of radiation that could penetrate several inches of aluminum, but not iron. The rays could be stored in some materials, for example, brick. Many eccentric claims followed. People could give rise to N radiation, but only in the absence of noise. There were also negative N rays, which allegedly had remarkable optical properties. When challenged about the incompatibility and apparent contradictions of some of the experimental results, Blondlot explained that N rays did not obey the usual laws of physics. Eventually, controlled experiments performed in Germany showed the phenomenon for what it was: a case of self-deception.

About 150 years earlier there had been the famous water-into-earth controversy, which bears a remarkable resemblance to polywater. It was claimed that when carefully distilled water was boiled away in a glass vessel, it left behind a mineral residue. The amounts of solid recovered always lay just at the limit of detectability, and not every sample would yield a solid residue. The debate attracted much attention on the part of chemists, physicists, and mathematicians, among them some famous names: Lavoisier, Boyle, Newton, and Leibniz. Eventually, in the face of much resistance, Lavoisier showed that the solid residue originated from the vessel in which the water was boiled. However, the same volume of the journal in which he published his findings also contained a long account by a Swedish chemist claiming that he had converted water into earth. Even in 1830 the question was still open. Although Lavoisier nowadays is given the credit for finally disproving the Aristotelian theory that water and earth were interconvertible elements, his contemporaries were not at all convinced.

Measured against Langmuir's six criteria, polywater, like the other two examples cited, certainly exhibits the symptoms of pathological science. Never more than a few micrograms were available for any analytical or physical measurements, so that the experimental techniques had always to be pushed to the ultimate limits of instrumental sensitivity. The liquid could never be made to condense in all the capillary tubes of any given batch, but only in about 30 to 40 percent. The very existence of the newly discovered substance rested on the premise that some completely new and unknown type of chemical bond was involved. Doubts relating to the presence of impurities were brushed aside by statements that in the believer's laboratory all due care was taken to exclude impurities. Finally, the ratio of supporters to critics also closely followed the course predicted by Langmuir.

With the benefit of hindsight it is easy to write off polywater as an aberration of scientific progress, but all this was not nearly as obvious in 1968. At that time polywater had to be taken as seriously as any other purported new chemical compound. It is only proper that news of some extraordinary discovery should receive publicity in the scientific media. Special attention should be paid to such a discovery when it originates from a center of orthodoxy, such as the Institute of Physical Chemistry of the Soviet Academy of Sciences. It is also proper for journals to permit followup work to be published and to act as forum for enlightened discussion, at least for a year or two.

Whereas Langmuir discussed pathological science in terms of the phenomena observed and their explanation, John Ziman, in his address to the 1970 meeting of the British Association for the Advancement of Science, talked about it in terms of breakdowns in the social and psychological mechanisms of scientific research. The most important question about such a breakdown concerns the integrity of the researchers themselves. Ziman suggests that scientists can, and often do, maintain two very different sets of standards. They "will intrigue for political ends like any Jesuit, and can be as lordly as any hospital consultant in the control of their juniors. They can deceive their wives, fiddle their tax returns, drive drunkenly, live beyond their means, feed parking meters, beat their children, and otherwise behave as antisocially as anyone else when the occasion demands." Ziman goes on to say: "I have the very strong impression that these traits are repressed within research itself." If such a thesis is to be

taken at face value, then one cannot help but wonder why
this ability to maintain divergent sets of standards should be
confined to scientists. Perhaps Ziman, as an insider, was
overrating the normal level of professional integrity among
scientists. There are certainly cases on record of lawyers who
absconded with clients' money and politicians whose actions
were influenced by "gifts" or contributions to campaign
funds. Forgeries of documents and works of art are common.
Ziman suggests that for scientists the rewards of cheating are
too meager to make it worthwhile, and also that their training
in research is likely to have sharpened their critical faculties:
"A fierce and uncompromising honesty is one of the standard attributes of the so-called scientific attitude." On the
other hand, it can be argued that it is the fear of swift retribu-
tion that is responsible for the apparent integrity of scientists.
The "property" of the researcher is elusive—it is new knowledge, which can only be turned into "property" by its disclosure. It then becomes subject to policing through the critical
judgment of the peer group; reward or punishment follows
swiftly.

The history of science certainly confirms that examples of
large-scale cheating are rare, whatever the reasons. Cases
such as that of Piltdown Man serve to show that for a hoax to
be successful in the face of currently available analytical
methodology would require a veritable conspiracy, involving
many individuals. As regards polywater, we can certainly discount dishonesty for profit, which is to say, for the promise of
scientific glory.

There is, however, another incentive to dishonesty among
scholars. Scientists can take up, or be forced into, positions
in defense of theories that make it hard, even impossible, to
extricate themselves "with honor" once the weight of evidence mounts up against them. In such situations there have
been cases of forged experimental results. The *cause célèbre*
is the fraud of the "Midwife Toad," perpetrated in the
Vienna laboratory of the biologist Paul Kammerer in 1925. It
has never been clarified who committed the forgery of the

scientific evidence, although the motive was fairly clear. Kammerer had for many years supported the Lamarckian hypothesis that acquired characteristics could be inherited against the neo-Darwinists, who upheld the theory of natural selection and genetic mutations. The dispute became increasingly vitriolic, far surpassing in venom anything we witnessed in the polywater debate of the early 1970s. Kammerer was made the victim of a personal vendetta directed from Cambridge, and his scientific integrity was openly called into question. Arthur Koestler in his scientific detective story *The Case of the Midwife Toad* provides a fascinating account of the events that forced Kammerer into an untenable position. After years of innuendo that his experimental evidence was faked, Kammerer's downfall was brought about by a fairly crude forgery of which he had in all likelihood been completely unaware. He shot himself on the eve of taking up a chair at the University of Moscow, where his Lamarckian views were at that time politically favored. Koestler's analysis demonstrates how objective scientific debate can degenerate into an irrational persecution that, in turn, can lead to personal tragedy. Polywater had some of the ingredients of the Midwife Toad affair, although no accusation has ever been made of forged experimental results.

If we discount cheating, we must still consider the possibility of self-deception. This is a gray area familiar to most experimentalists; the wish to be objective is paired with the hope that the data will confirm one's thesis. Expressions such as "cooking the results," "massaging the data," and "wishful interpretation" are commonplace among scientists, theoreticians and experimentalists alike. It is hard to draw the borderline between conclusions based on observations made at the limit of sensitivity of some instrument or the human eye, on the one hand, and self-deception on the other. Yet only the latter could be classified as pathological science. Most experiments are performed because the scientist thinks he already knows what the outcome will be, or at least what it ought to be, to fit some preconceived hypothesis or to disprove a theory advanced by a competitor. The label "self-deception'' or ''pathological'' can be applied only if the experimenter has not taken appropriate precautions and has not carried out the necessary checks on his results. Ideally, any set of experiments, if repeated by an independent investigator, should produce identical results, but this is frequently not the case, especially when the laboratory procedures require a high degree of skill. In such a situation it may be hard to decide who is right, and subjective judgments are unavoidable.

Modern developments in statistics have led to improved methods for designing experiments and analyzing results, but the elements of human judgment can never be wholly excluded. In 1936, for example, it was recognized that Mendel's experimental results—which formed one of the main foundations of the case of the neo-Darwinists who persecuted Kammerer in the 1920s—had been faked. However, since Mendel's laws have been shown to be correct, not much attention has been paid to the faked statistics. According to Koestler, the forgeries quite likely arose from self-deception on the part of Mendel's assistants, who knew the elements of his hypothesis and tried to be helpful by tilting the evidence in favor of their master's expectations.

That self-deception was one of the ingredients of the polywater affair, especially after the bandwagon had begun to roll, cannot be denied. Therefore, on this count, the charge of pathological science must be taken seriously.

I have noted that undue polarization of opinion within the scientific community leads to pressures to perform uncritical experimental work that can, once the dust has settled, be shown to have been of unacceptable standards. During the brief history of polywater such polarization became the order of the day. Normally accepted conventions of impersonal expression went by the board. Even when it was all over, the polarization remained; the ''polywater supporter'' label is still a hard one to live down. This must be in part due to the fact that the drama was not played out behind the closed doors of laboratories and in the scientific journals. It received more than its fair share of press coverage, and quite a number of the people involved could not resist the temptation of giving press interviews and allowing half-considered views to be printed. In the words of economist Paul Samuelson, they were unable to distinguish "the gold of scientific fame from the brass of popular celebrity." They should have known that he who submits to a press interview often has to pay dearly for the publicity. He has no control over what is printed. His statements may be distorted, quoted out of context, and given a completely different meaning from that intended. All this is considered a fair price to pay for news value. Several scientists thus became household names, and in the aftermath they were not forgiven for it. It is certain that had Lippincott not died prematurely, his name would have remained linked with polywater for the remainder of his active career and he would have suffered a good deal of ridicule—not so much because his spectroscopy turned out to be wrong, but because of the publicity which he received.

Such personal factors as I have described, which played an important role in the polywater affair, are of course ingredients of most scientific developments. Witness, for instance, James Watson's *The Double Helix,* in which he describes one of the most significant discoveries of modern times, one that led to the award of several Nobel prizes. The picture Watson paints differs radically from the accounts to be found in the textbooks on molecular biology, because he dwells on the everyday lives of the researchers. It is quite clear that even when the work was going well, personal dislikes, competition, secretiveness, and misleading the "competition" were the order of the day, and the drive "to get there first" was strong if not dominant. All this is quite in line with Ziman's evaluation of the human face of scientific research. Can it be labeled pathological? Indeed, it has been said that Watson's revelations first demonstrated to the public at large that scientists are human. Francis Bacon, in his *Advancement of Learning,* has put it neatly: "We are much beholden to

Machiavel and others, that write what men do, and not what they ought to do.''

Where, then, is the borderline between what is normal and acceptable and what is pathological? Perhaps we can look for it in the reaction by the scientific world to revelations such as those contained in *The Double Helix.* On the whole this reaction was negative, marked by recriminations and personal invective. Had Watson unwittingly succeeded in demonstrating not only that scientists are human, but that on occasions they are too human? Any fair-minded practicing scientist must admit that the atmosphere of the everyday life in a research laboratory described by Watson has a ring of familiarity about it. The reaction showed one facet of the scientist's emotional makeup: a dislike of the rough and tumble of popular debate, which he regards as ''uninformed.'' In this he is no different from members of other professions (witness the outcry from the lawyers every time the subject of legal reform is mentioned by a newspaper or a politician).

What is particularly hard to forgive is when a member of the profession turns ''traitor'' and stoops to write a popular account of some aspect of the profession, its institutions, or its members. In the introduction to the British edition of Daniel S. Greenberg's classic *The Politics of Pure Science in the U.S.A.* Robin Clarke describes the reception of the book by the American scientific community when it was first published in 1967. Joel A. Snow, reviewing the book in the *Bulletin of the Atomic Scientists,* considered it ''. . . in easy contempt of the conventions of politeness in print which are customary in the scientific world.'' Frank T. McClure wrote in *Science* (where Greenberg had been news editor) that ''the overall effect is to demean, and few men or institutions went into this book but came out poorer.'' In view of such attitudes, the animosities that developed during the polywater affair are perhaps easier to understand. The degeneration of normal scientific debate into something that was later judged as pathological may have been sparked off by the intervention of the media and the readiness of some individuals to give off-the-cuff opinions and views, not only about polywater but also about the personalities of its researchers.

Both the Langmuir and the Ziman descriptions of pathology in science really concern deviant behavior among practitioners rather than deviant science. In other words, there is really no such thing as sociology of science; what we are concerned with is the sociology of scientists. Ziman's analysis suggests that, as a group of professionals, scientists are little different from other such groups: They are identifiable by a code of professional standards, by certain patterns of communications, by institutionalized training and qualification procedures, by a strong professional hierarchy, and by a rather individualistic reward system. However, about the discipline of science divorced from the human element, sociology has little to say in comparison with the profound contributions of philosophy. Let us remember, therefore, that in discussing any atypical sociological aspects of polywater our concern is primarily with the manner in which this discovery was handled by those who became involved with it.

The only sociological commentary on the polywater episode so far forms part of a Harvard University M.A. thesis by Simon Schaffer, who reached his conclusions after sifting the evidence and interviewing several of the "polywater personalities" of the early 1970s. Schaffer lists four facets of the polywater work that could be construed as pathological signs. Characteristically, they are rather different from those proposed in more general terms by Langmuir. They are the hostile attitudes shown by the scientific establishment, which in turn affected the financing of research; the quality and quantity of press exposure; the lack of communication and secretiveness adopted by the researchers; and the amount of hasty and shoddy research.

In examining the behavior of the scientific community, we must draw a line between those researchers who were personally involved, and whose reputations were therefore at stake, and those whose activities consisted of comment, advice, and criticism. This latter group included members of the scientific establishment who, at that time, had the ear of those holding the purse strings. They also had easy access to the media and were thus in a position to affect the climate in which the research was carried out.

It hardly needs emphasizing that scientists as a group are very sensitive to the views of their peers, more so than they are to the esteem in which they are held by the public at large. After all, recognition by the peer group is the currency of their reward system. One might suggest that the indifference scientists at times show to the general public is ill advised, as they rely on vast sums of public money. In more recent years this insight has led to public-relations exercises in which attempts are made to explain the "relevance" of major research undertakings. Scientists are of course not alone in relying for their immediate livelihood on public funds, but they are unique in modern society in also having to rely on public money for buildings, hardware, and other facilities before they can begin to practice their craft. One can read continuous debates about the return science provides, or how this can be measured. Comparisons are sometimes drawn between scientists and the high priests of old or the builders of great cathedrals in the Middle Ages or the Renaissance, when the public was persuaded to part with large sums of money in return for benefits—sometimes intangible—to be derived at some later date or even in another life. In other words, the support of major scientific endeavors requires an act of faith. However, in the case of polywater, the development was not likely to cost very much and the potential industrial and economic benefits were immense. Even the *Wall Street Journal* had been convinced of this at one stage. Polywater research was therefore an enterprise that could be actively supported without too much trouble; at least, so it seemed in the early days.

Regarding the growth and makeup of the polywater research community and its relations with the scientific establishment, mainly in the United States, we can see that it differed from other recognizable research groups by its lack of homogeneity. Polywater work, originally the preserve of the surface chemists, soon attracted physicists, chemists, and life scientists of many different persuasions and backgrounds. Intrinsically such a multidisciplinary approach to the solution of a major problem can have much to commend it, but in the particular case of polywater it bore signs of a scramble rather than of organic growth—witness the applicability of the mathematics of epidemics to the development and decline of research on the subject.

The only available means of identifying members of the polywater community is through the published scientific abstracts, where their names are linked with the keyword "polywater." The group bears some resemblance to the army of mercenaries that roamed across Europe during the Thirty Years' War. These soldiers of fortune owed allegiance to the prince or general who paid them, and they were quite ready to change sides should the occasion demand it. Undoubtedly a considerable number of individuals had their own personal religious convictions and were loyal to a cause, but it was hard to differentiate them from their comrades on the battlefield. So it was with polywater: Some of the protagonists and antagonists were very likely motivated completely by the scientific challenge. Their contributions, even where they have eventually turned out to be wrong, bear all the marks of thoughtfulness in their conceptions and of high scientific standards in the prosecution; but these were exceptions.

This again raises the issue of priority. Research work must be seen to be original, and originality is demonstrated by priority in publication. On the other hand, scientists are taught to be personally modest and to assess correctly the true value of their contributions to knowledge. Thus, one is encouraged— almost incited—to publicize one's originality, and yet convention demands that one admit how little has really been accomplished, and how one cannot even take credit for all the work one presents. This admission is usually expressed by the listing of multiple authors in even the shortest communications, acknowledgments of gratitude to a host of individuals, and lengthy bibliographies designed to honor correct priority. In the contest between the drive for originality and a learned humility, it is the former which usually wins, with all the attendant evils.

Schaffer has identified the twin characteristics of hasty and shoddy research and lack of communication and secretiveness with pathological science. This can only be a matter of degree, because scientific research is firmly based on competition. When the basic financial needs of the scientist are secured, job satisfaction can then be equated with professional recognition, and this in turn depends on "originality." It has often been stated that the "search for truth" is in itself a motive for scientific endeavor, but it is rarely the prime motive. If ever the delicate balance between making a living, searching for the truth, and obtaining the approbation of one's peers is upset, then there is the danger of deviant science. The signs are familiar to most professional scientists: The race to publish gathers momentum, speculation takes the place of information, competitors are accused of plagiarism or selective use of evidence, and, commonest of all, publications take on a partisan character (they might for instance omit any mention of related work performed by others). Thus, intense competition evades and sometimes violates the accepted professional norms.

The history of polywater provides clear evidence of questionable publication practices. One of its distinguishing features is the proliferation of "quickie" publications, for instance in *Nature* and *Science.* The majority of those who got involved in the affair published no more than one such letter in one of these two journals. The pressure was great—in the general rush for priority there was not enough time to perform enough experiments and think about the results before committing them to paper. There can hardly be an active scientist who has not at some time been under such pressure to reach a certain stage in his work, either because his grant is running out, or because he is committed to present a paper on work that is yet to be completed, or because he has heard over the grapevine that a competitor is just about to preempt him. Even with tight deadlines, however, the quality of the work must remain of paramount importance, since indiscriminate corner-cutting always leads to trouble. Such pressures usually develop when the scientist has already been active in his field for some considerable time and therefore has a personal stake in its future development, but this was not so in the case of polywater. The army of researchers grew from just a handful in 1968 to well over two hundred by 1972, to which must be added many more whose names were not actually cited as authors in publications. Initially, few of them had a stake in the future of polywater; therefore the only conceivable motive for quick and short publications of quick and short experiments was the fear that somebody else might get there first. Whatever unjustified accusations may have been leveled against polywater researchers, the charge of jumping on the bandwagon is a hard one to refute in the light of the printed evidence.

The charge of secretiveness and lack of communication is much harder to substantiate. The polywater activists had an unusually active grapevine. In fact, results had a habit of being communicated by telephone. On that score there was no secretiveness; everybody knew pretty well what everybody else was doing and how far he had gotten. The uneasiness and insecurity of the polywater researchers is best illustrated by the lack of exchange of samples for study and the lack of true collaboration between laboratories of different persuasions after the initial joint venture between Bellamy and Lippincott. Had such an exchange taken place, then independent investigations on common samples would soon have shown up the possible effects of impurities on the recorded spectra. As it was, any such doubts were always countered by the claim that the doubter had not yet perfected the proper techniques for the preparation of pure polywater. Thus, the uncertainties could have been resolved much sooner, but then scientific research does not and cannot always proceed by the most economic pathways, especially without the benefit of hindsight.

The attitudes adopted by members of the scientific establishment, and the resulting developments, have been identified by Schaffer as another facet of pathology in the polywater affair.

Like other professions, science has its institutions and its hierarchy. It could hardly be otherwise, because democracy and equality have no place in a profession. The upper echelons of the hierarchy form the so-called establishment. By and large this consists of the collective membership of the various national academies of science. This membership in turn is made up of prominent academics, with a very thin sprinkling from industry or government scientific institutions. Members are ostensibly chosen for the excellence of their work and for their productivity. These attributes are assessed by the number of papers published and their quality.

The functions of the scientific establishment differ from country to country and from discipline to discipline. Paramount is the maintenance of high professional standards, and this is responsible for the conservatism and orthodoxy with which the establishment is usually associated. While it is of course right and proper that the establishment should not too readily support views which smack of scientific heresy, its members must always be aware of their power to make and break the careers and prospects of the younger generation. The correct assignment of credit to a junior scientist is not always treated as scrupulously as it might be, and so it was with polywater. Few chemists were ever aware of the existence, let alone the work, of Fedyakin. The name of Deryagin became associated with anomalous water fairly well from the beginning, and Deryagin was very visibly a member of the establishment. One can argue that had he not taken an interest in the project of the unknown Fedyakin, anomalous water would have died there and then. However, not only did Deryagin take an interest, but he appropriated the discovery— a common enough occurrence in the annals of science.

By and large, the scientific elite was not only hostile but intolerant to polywater. Only a few of its members went so far as to comment on polywater research in the scientific literature, but where they did their comments are tinged with sarcasm and thinly disguised personal invective. Most of the elder statesmen of American science who had any view on the subject made their feelings known more informally, such as during discussion periods at conferences. I was to witness many negative remarks by respected individuals who could have reserved judgment at the time. After all, a fair number of those who (wisely or unwisely) had decided to get into polywater research were themselves highly reputable and well-known scientists with orthodox records. For this reason alone, members of the establishment should have exercised a degree of charity. Their lack of tolerance conributed in no small measure to the polarization of opinion that became so apparent immediately after the Lehigh meeting of 1970. A. T. Huxley once commented that in science ''merit alone is very little good; it must be backed by tact and knowledge of the world to do much good.'' This is unfortunate, but true. However, there are occasions when even tact and knowledge are not enough to gain a hearing for unorthodox ideas unfamiliar to members of the scientific elite, whatever their intrinsic merit. There are well-documented instances of a major discovery being thrown out by some august Academy of Sciences, only to gain acceptance several years later when put forward in a more ''acceptable'' manner or by a more acceptable member of the scientific community. An important example of such conduct is the German Academy of Sciences' failure to accept Polanyi's novel ideas on the adsorption of gases on solid surfaces. This caused Polanyi much suffering, and one can only guess at his feelings when, several years later, Irving Langmuir rediscovered (quite independently and by a different method) the phenomenon of adsorption.

One of the most serious charges that can be leveled against a scientist by his peers is that of unscientific conduct. Such charges were certainly made in the polywater affair, particularly in the later days. The "experts" were once again proved right, after they had arrogantly asserted all along that polywater did not and could not exist. These assertions were based more on hunch than on reasoned argument or study of the evidence. Even those who had worked hard and persistently to disprove Deryagin's claims were galled by the high-handed way in which members of the establishment dismissed the experimental evidence and yet turned out to be right in the end. The members of the elite are far from infallible, but they are conservative, and in science conservatism pays off more often than not.

If in the early days the polywater researchers had only had to face disapproval and personal criticism from the elite, they would have weathered it and carried on with their work. But closely linked with the sentiments of members of the establishment was their influence over the financing of research. Science is now funded almost completely from public sources through various government-appointed agencies. Over the past decade the mechanisms whereby money finds its way to the individual scientist have been revised and refined several times so as to make the process as fair and open as possible, at least in the United States. Inevitably this has also resulted in longer delays and a more complex administrative machinery of committees, panels, working groups, and reviews. However, in the late 1960s the power of the elite was still very strong. Its members or nominees staffed the various councils, committees, subcommittees, and working groups that made up the machinery of science funding, and it was not uncommon that unconventional albeit highly original research proposals were refused financial support. At the time no reasons had to be given for such re-

186

fusals, and the hapless applicant could be left without guidance as to the merits or shortcomings of his ideas and plans.

Rejected research proposals make interesting reading, especially when several years later the plans originally put forward have been successfully realized and the results published with general acclaim, usually by somebody else. The sociologists have recognized the sometimes incomprehensible resistance to scientific discoveries and have speculated whether such resistance helps or hinders the progress of science. A particularly pernicious device was used to discredit Leland Allen. In the autumn of 1970 he submitted a research proposal, completely unconnected with polywater, to the Air Force Office of Scientific Research. Using the customary cloak of anonymity, one of the referees had this to say: "Allen modestly points out that he has been productive. However, when a principal investigator publishes a paper such as 'A Theory of Anomalous Water,' one wonders whether the exposure given AFOSR is favorable or unfavorable. I presume that AFOSR is more interested in supporting credible research than it is having its name spelled right on incredible research. Allen's paper, and some of his other papers in the past, may be described as incredible." Another referee commented: "All of chemistry, including that part of great relevance to the Air Force, will be revolutionized by this type of work. Further, Allen's record is exceptional, both for quality and quantity." With a diversity of opinions of this magnitude, it is hard to assess the quality of a proposal (we do, however, learn something about the personalities of the referees).

The funds officially provided by the various governmental sources for polywater research are a matter of public record. In the beginning, pressure from the U.S. Office of Naval Research was responsible for the funding of various feasibility studies. This led to the allocation of funds to several American universities by various agencies, mainly those with a stake in defense. Such grants were awarded primarily for the development of methods to produce polywater in large quantities. Before long, adverse comments by members of the establishment—frequently off the record—led to a drying up of funds.

The fickleness of the funding agencies is illustrated by Leland Allen's experiences. In the fall of 1969 he was approached by two Department of Defense agencies and invited to submit large polywater proposals immediately. For various reasons he declined. In February 1970 he asked informally for a small grant to pay for computer time for his polywater calculations, and was told that there was no longer any interest, since several distinguished consultants had told them that ''they would look silly supporting such nonsense.'' Also, during the same month, February 1970, a panel of the National Science Foundation rejected all of ten polywater research proposals, because one of its members had heard that a ''reputable scientist'' had proved conclusively that polywater did not exist. These are not isolated examples of the way in which the decision-making process operated. However, the heavy hand of the elite did not show itself in the suppression, by adverse refereeing, of many of the research reports on polywater. One can only speculate either that the refereeing was superficial or that on occasions the editors overrode the recommendations of referees in order to promote further discussion, or to provide a forum for unorthodox views.

Despite the drying up of financial support, the bulk of polywater research was performed during 1970 and early 1971. We must therefore conclude that most of this work was funded from resources originally earmarked for other purposes, a conclusion that has been confirmed by several of my correspondents. It is just one example of the sophisticated bootlegging systems that have been developed by scientists to mitigate the effects of sudden fluctuations in the funding of research. In this connection Simon Schaffer has recorded comments by scientists that the U.S. Department of Defense would ''put money into far-out ideas at the earliest report and then pull out just as quickly''—a practice hardly conducive to a well-planned research effort. This is a criticism often leveled against the methods by which industry finances research. It has been suggested that, had such haphazard methods been applied in the very early days of technical development, we would now possess the most superior stone axes but little else.

Perhaps one of the main factors that distorted the treatment afforded to polywater work by the U.S. government funding agencies was the fact that it was discovered in the Soviet Union. During the early 1960s, Sputnik was fresh in the minds of those responsible for directing the course of American science and technology. Now the Russians had an undisputed lead of several years with polywater. This lead had to be narrowed, if not eliminated, in as short a time as possible, and little thought was therefore devoted to a methodical evaluation of the available data. When government agencies reconsidered and abruptly discontinued their support, the scientists who not long ago had been encouraged and had meanwhile become committed to their polywater projects felt let down and cheated. To quote Schaffer: ''The elite saw polywater as bad science, while the researchers saw the behavior of the elite as corrupt science.'' Here then were some of the ingredients of pathological science.

The involvement of the media in the polywater debate, identified by Simon Schaffer as a malignant influence, probably makes polywater unique in the catalog of ''the science of things that aren't so.'' The rapidly increasing professional tension that is apparent from a perusal of the scientific literature of the day was fueled by the popular press. The tension was increased yet further by some of the scientists who allowed themselves to be interviewed and quoted (or perhaps misquoted) in the popular science journals or, worse yet, in the mass media. More often than not the tone of such reported quotations exaggerated the existing differences of opinion. For instance, the group at Birkbeck College, themselves among the first ''activists'' in 1967, suggested several years later that Deryagin's discovery had been taken seriously not so much because of its intrinsic merit but mostly because of Deryagin's professional standing as a member of the elite. They then expressed the view that "much of the published work since then is careless, inconclusive, and often misleading, if not wholly conjectural." J. J. Bikerman, noted for his contributions to surface chemistry, had been silent on the subject of polywater for all of six years, until after Deryagin admitted his change of mind in 1973. At that point Bikerman felt impelled to comment that Deryagin's recantation was not convincing and that his previous observations could not be accounted for by small amounts of impurities, but that rather "the protracted error immortalised in scores of publications can be due only to psychological rather than to chemical causes." These are but two of scores of similar comments which are to be found in the *scientific* literature of the day.

Time and again the proponents of polywater were "accused" of practices that are in fact common and acceptable to professional scientists, such as the publication of ideas and hypotheses not yet thoroughly tested and confirmed—a fairly common practice, as illustrated in *The Double Helix.* This readiness to judge is an aspect of the competition among scientists. The degree to which "premature" publication is acceptable is uncertain and varies from issue to issue. At the height of the controversy, several months before the Lehigh symposium, Lawrence B. Chase quoted Leland Allen in the Princeton *Alumni Weekly:* ". . . there is too much pressure which mediates against risk, and in favor of authority and established theories. . . . The sociology of science generally forces you to do well-defined, solid work, *not* to take leaps, not to take the chance of being wrong." Characteristically, Kollman, Allen's graduate student, realized that because of his status he "didn't have to face the kinds of skepticism and disbelief from professional colleagues [that Allen had]." He also judged that ". . . as a graduate student, I don't stand to lose as much if we turn out completely wrong.'' The moral seems to be that radical hypotheses are only permissible if the results turn out to be correct and acceptable at the time.

The admissibility of a change of mind, especially by a leading figure, is also open to debate. The manner in which Deryagin wrote off ten years of work and publicity in two sentences was not easy to accept for those who had followed him, and led to some bitter criticism. Deryagin, however, was geographically inaccessible. Scapegoats had to be found closer at hand. Leland Allen, in particular, was made to shoulder much of the responsibility. Few of his contemporaries were willing to take at face value his change of mind, although there is no evidence that this was arrived at other than by the gradual persuasion that his earlier calculations had led him to mistaken conclusions. A bad case of self-deception? Perhaps, but it is more likely that Allen's only mistake was to believe and claim that his theoretical techniques were so refined that they could discriminate between quite similar types of chemical bonds. The literature is full of such misinterpretations of experimental data, made in good faith on the basis of the techniques and information available at the time of writing.

Michael Faraday once expressed the hope that 50 years after his death nothing he had ever written would still be considered true. Most scientists of today lack Faraday's modesty; they like to believe that everything they commit to paper will endure as truth forever. Genuine changes of heart or mind are therefore suspect, and this may be why Allen drew upon himself the wrath of some members of the scientific community (in particular, that of the elite). A number of the elder statesmen of science who were quick to condemn Allen have themselves invested much time and effort in defending theories developed many years ago that in their day were original and ingenious but have been overtaken by events, a fact known to all but the defenders. There should be nothing shameful or dishonest in a change of mind genuinely arrived at on the basis of new evidence. It is here that intellectual pursuits must differ from politics: It is a loss of face for a politician ever to admit to a mistake, but it should not be unforgivable for a scholar.

On the credit side it must be emphasized that, although the polywater debate led to acrimony and animosity among scientists, after its eventual decline the former supporters were not subjected to any witch hunt, at least not in the United States or Britain. Most of those whose names were associated with polywater are still active, and their careers do not appear to have suffered unduly as a result of past activities.

In the Soviet Union events took a slightly different turn. On the face of it, Deryagin did not suffer as a result of his leading role. However, he has not been awarded full membership of the Academy of Sciences, although in 1970 there was talk of a Nobel prize. The real discoverer of anomalous water—Fedyakin—disappeared from the scene and presumably went back to the provinces; his name has not appeared in the scientific literature since the zenith of polywater. The fate of Zhelezhny, another of Deryagin's junior colleagues, is a classic in the "Catch 22" tradition. After taking part in the early experiments, he was worried about the possibilities of impurities in the anomalous water and, on his own initiative, submitted a sample to an East German spectroscopist for analysis. The results indicated appreciable levels of impurities, news which Zhelezhny brought to the attention of his superiors. Deryagin did not consider these experimental findings worthy of note in his reports. However, Zhelezhny's name was omitted from all further papers describing research on anomalous water. Several years later, after Deryagin himself had become convinced of the role played by impurities in changing the physical properties of water to those of anomalous water, Zhelezhny was held responsible for not having done his job properly. He disappeared from his post at the Physical Chemistry Institute in Moscow and is now employed at the Institute of Soil Science in Leningrad.

Something was certainly wrong—quite apart from the substance polywater itself—in the manner in which the affair developed. I have tried to separate the ingredients that through their complex interplay made the polywater research so atypical of scientific developments. In the case of polywater, the immune system of the body scientific failed to respond effectively to the external and internal threats of infection. These threats were manifold: the methodology adopted by the researchers, their communications network, and their motivation. These internal factors, however, were conditioned largely by the environment—publicity, opinions of fellow scientists, attitudes of funding agencies, career prospects, political considerations, and maybe even foreign policy. Without doubt, the most powerful single element responsible for turning polywater research into pathological science was the involvement of the mass media, which thrive on dispute and confrontation.

With a few aberrations, the progress of science since it became an important factor in public affairs has been such that the profession has had little to be ashamed of. Anomalies have been dealt with quietly and expeditiously within the scientific community, with professional decorum and by accepted procedures. The one distinguishing feature of the polywater debate was the publicity it suffered. The usually slowish process of scientific publication, filtered by refereeing, could be short-circuited by a press interview. This eventually played havoc with the conduct of the research, its communication, its quality, its funding, and its apparent importance. Scientists became involved who, in different circumstances, would never even have heard of polywater until after it had been disposed of as a debatable issue.

It is impossible to pass a verdict on the polywater affair in terms of ''guilty or not guilty,'' though some have seen fit to do so. Rather, we should build the experience into the general fabric of scientific and sociological knowledge and apply the results and conclusions to future scientific developments. That, after all, is the way in which science has progressed so

successfully in the past. In the meantime, we have been reminded that the scientist does not live and work in a vacuum,
but is conditioned by and responds to the influence of his environment. In the late 1960s these influences were such as
to encourage developments that led to the type of activity
exemplified by polywater. Since then the environment of the
scientist has changed dramatically; he is no longer the dar-
ling of society, and he has to justify his plans and achievements not only to his peers but to all manner of other special
interest groups. It is hard to predict how the scientific community would respond to another Fedyakin-type discovery in
the 1980s, but probably the science fiction that might form
the basis of such a discovery has already been written.

Vonnegut's statement that ''pure research men work on
what fascinates them, not on what fascinates other people''
is as true as it ever was, despite much effort on the part of
the planners. Discoveries cannot be legislated; the progress
of knowledge takes place in a random fashion. There can be
no insurance against another polywater, and on balance such
a lack of insurance is likely to benefit the scientific community and the society it should serve.

Epilogue

From *Science,* 27 October 1978:

Twin brothers at a relatively obscure research institute in Kazakhstan have identified [a] form of water that appears to have a greater biological activity than ordinary tap water. . . . Soviet scientists have known for some time that fresh meltwater has the capacity to stimulate some biological processes. It has been theorized that the meltwater retains some of the order that is characteristic of frozen water and this increased order alters vital reaction rates within cells; some American investigators have theorized, similarly, that water within cells is more highly ordered than ordinary water and that this increased order is essential to proper control of enzymes.

Just as I finished the manuscript of this book, the above appeared in *Science;* hard to believe, but true. This time, however, the "bioactive" water is easy to prepare in large quantities. All that is required is a milk-pasteurizing unit. It is claimed to possess some wonderful properties: Cotton plants grown from seeds that have been soaked in bioactive water yield 10 to 12 percent more cotton than ordinary seeds; the same is true for other crops, especially sugar beets. The water increases the hemoglobin level in the blood of experimental animals and causes cattle to gain weight. Concrete prepared with it is 8 to 10 percent stronger than concrete prepared with ordinary water.

The report ends with the words "Soviet investigators predict that a variety of other uses will soon be found for this 'magical' material." Only the future will tell what lessons we have learned from polywater.

Bibliography

This chapter-by-chapter bibliography is by no means comprehensive, nor is it confined to the subject of polywater. It lists the significant scientific publications describing polywater research, related stories in the popular press, and publications on peripheral topics discussed.

Two extensive polywater bibliographies are in existence:

Allen, L. C. "An Annotated Bibliography for Anomalous Water." *Journal of Colloid and Interface Science* 36 (1971):554–561.

Gingold, M. P. "L'eau dite Anormale: Revue Generale." *Bulletin de la Société Chimque de France* No. 5 (1973):1629.

Chapter 1

Franks, F. "The Unique Chemical." *Chemistry in Britain* 12 (1976):278–281.

Franks, F. "Solvent Mediated Influences on Conformation and Activity of Proteins." In *Characterization of Protein Conformation and Function,* edited by F. Franks. London: Symposium Press, 1979.

Franks, F., ed. *Water—A Comprehensive Treatise.* 6 vols. New York: Plenum, 1972–1979.

House, C. R. *Water Transport in Cells and Tissues.* London: Arnold, 1974.

Vonnegut, K. *Cat's Cradle.* London: Gollancz, 1952.

Chapter 2

Deryagin, B. V. "Effect of Lyophile Surfaces on the Properties of Boundary Liquid Films." *Discussions of the Faraday Society* 42 (1966):109–119. (See also pp. 134–140.)

Deryagin, B. V., Churaev, N. V., Fedyakin, N. N., Talayev, M. V., and Yershova, I. G. "The Modified State of Water and Other Liquids." *Bulletin of the Academy of Sciences of the USSR* (Izv. Akad. Nauk SSSR, Ser. Khim.), no. 10 (1967):2178–2187.

Fedyakin, N. N. "Change in the Structure of Water during Condensation in Capillaries." *Kolloid Zhurnal* 24 (1962):497 [*Colloid Journal USSR* 24 (1962):425–430.

Hazelwood, C. F. (ed.). *Physico-chemical State of Ions and Water in Living Tissues and Model Systems. Annals of New York Academy of Sciences* 204 (1973):1–631.

Chapter 3

Bangham, A. D., and Bangham, D. R. "Very Long-range Structuring of Liquids, including Water, at Solid Surfaces." *Nature* 219 (1968):1151–1152.

Bangham, D. H., Mossalam, S., and Saweris, Z. "Visible Adsorbed Films and the Spreading of Liquid Drops at Interfaces." *Nature* 140 (1937):237–238.

Burton, R. A. "Polymerized Water?" U.S. Office of Naval Research European Science Notes, 1968, no. 22-6.

Burton, R. A. "Moscow Visit." U.S. Office of Naval Research European Science Notes, 1969, no. 23-3.

Burton, R. A. "Orthowater Black-out?" U.S. Office of Naval Research European Science Notes, 1969, no. 23-1.

Burton, R. A. "Anomalous Water Round Up." U.S. Office of Naval Research Technical Report, 1969, no. R33-69.

Medvedev, Z. *The Medvedev Papers: Fruitful Meetings Between Scientists of the World.* London: Macmillan, 1971.

Chapter 4

(Anonymous.) "Enter 'Polywater'—Amid Alarums and Excursions." *New Scientist,* July 10, 1969, p. 55.

(Anonymous.) "Scientists Tell of a New Water." *New York Times,* September 12, 1969, p. 46.

(Anonymous.) "Polywater." *New York Times,* September 22, 1969, p. 32.

(Anonymous.) "Polywater." *Scientific American* 221 (1969):90–95.

Bellamy, L. J., Osborn, A. R., Lippincott, E. R., and Bandy, A. R. "Studies of the Molecular Structure and Spectra of Anomalous Water." *Chemistry and Industry* (1969):686.

Bernal, J. D., Barnes, P., Cherry, I. A., Finney, J. L., Everett, D. H., Haynes, J. M., and McElroy, P. J. " 'Anomalous' Water." *Nature* 224 (1969):393–394.

Bolander, B. W., Kassner, J. L., and Zung, J. T. "Cluster Structure of the Anomalous Liquid Water." *Nature* 221 (1969):1233.

Cherkin, A. " 'Anomalous' Water: a Silica Dispersion?" *Nature* 224 (1969):1293.

Donahoe, F. J. " 'Anomalous' Water." *Nature* 224 (1969):198.

Finney, J. L. "Polymerized Water: Is It or Isn't It?" *Chemistry* 42 (1969):20.

Lear, J. "The Water that Won't Freeze." *Saturday Review,* September 6, 1969.

Lenihan, J. "Polywater and the Stuff from Taps." *Guardian,* October 16, 1969.

Pothier, R. "Miami Scientific Team Creates Mysterious New Form of Water." *Miami Herald,* July 30, 1969.

Willis, E., Rennie, G. K., Smart, C., and Pethica, B. A. " 'Anomalous' Water." *Nature* 222 (1969):159.

Chapter 5

Allen, L. C., and Kollman, P. A. "A Theory of Anomalous Water." *Science* 167 (1970):1443–1454.

Allen, L. C., and Kollman, P. A. "What Can Theory Say about the Existence and Properties of Anomalous Water?" *Journal of Colloid and Interface Science* 36 (1971):469–482.

Chua, K. S. "Structure of Anomalous Water and its Mechanism." *Nature* 227 (1970):834–836.

Cherry, I., Barnes, P., and Fullman, J. "Comment on K. S. Chua, 'Structure of Anomalous Water and its Mechanism.' " *Nature* 228 (1970):590–591.

Chua, K. S. "Reply to Cherry, Barnes, and Fullman." *Nature* 230 (1971):379.

Dannenberg, J. J. "Predictive Molecular Orbital Calculations in Organic Chemistry." *Angewandte Chemie* (international ed.) 15 (1976):519–525.

Donohue, J. "Structure of Polywater." *Science* 166 (1969):1000–1002.

Everett, D. H., Haynes, J. M., and McElroy, P. J. "Colligative Properties of Anomalous Water." *Nature* 226 (1970):1033–1037.

Petsko, G. "The Search for Polywater" and Chase, L. B. "The Sociology." Princeton *Alumni Weekly,* March 10, 1970, pp. 6–15.

Chapter 6

Allen and Kollman. "What Can Theory Say . . . ?"

(Anonymous.) "Can He Prove It's Water?" *Chemical Week,* July 1, 1970.

(Anonymous.) "Polywater Existence Still Unsettled." *Chemical & Engineering News,* July 13, 1970, pp. 29–35.

(Anonymous.) "Polywater Theory Rebutted." London *Times,* July 14, 1970.

Bancroft, G. "Strange 'Polywater' is Quarry of International Science Search." *Morning Call* (Allentown, Pennsylvania, June 22, 1970.

Blakeslee, S. "Scientist Says Mystery of Polywater has been Solved: Russian's Test Samples Contained Sweat." *New York Times,* September 27, 1970.

Braud, D. "Weird Water: Debate Over Mysterious Fluid Splits Scientific World." *Wall Street Journal,* July 21, 1970.

Deryagin, B. V. "Superdense Water." *Scientific American* 223 (1970):52–71.

Donahoe, F. J. "Is Venus a Polywater Planet?" *Icarus* 12 (1970):424–430.

Lippincott, E. R., Cessac, G. L., Stromberg, R. R., and Grant, W. H. "Polywater—A Search for Alternative Explanations." *Journal of Colloid and Interface Science* 36 (1971):443–460.

Middlehurst, J., and Fisher, L. R. "A New Polywater." *Nature* 227 (1970):57.

Rossotti, H. S. "Water: How Anomalous Can It Get?" *Journal of Inorganic Nuclear Chemistry* 33 (1971):2037–2042.

Rousseau, D. L. "An Alternative Explanation for Polywater." *Journal of Colloid and Interface Science* 36 (1971):434–442.

Rousseau, D. L., and Porto, S. P. S. "Polywater: Polymer or Artifact?" *Science* 167 (1970):1715–1718.

Sullivan, W. "Researchers Cast Doubt on Finding That Water Can be Converted to a Dense Vaseline-Like Form." *New York Times,* April 2, 1970.

Symons, M. "Scientists' New Answer to Mystery Polywater." Sydney *Morning Herald,* July 1970.

Symons, M. "Polywater? There's No Such Thing, Says C.S.I.R.O. Man." Sydney *Morning Herald,* July 1970.

Chapter 7

Allen, L. C. "Polywater." *Science* 173 (1971):1252.

Allen, L. C. "The Rise and Fall of Polywater." *New Scientist,* August 16, 1973, pp. 376—380.

(Anonymous.) "Polywater's End." *New York Times,* July 28, 1973.

(Anonymous.) "Next Slide, Please." *Nature* 272 (1978):743.

Baglow, G., and Bottle, R. T. "Rate of Publication of British Chemists." *Chemistry in Britain* 15 (1978): 138—141.

Barnes, P., Cherry, I., Finney, J. L., and Petersen, S. "Polywater and Polypollutants." *Nature* 230 (1971):31—33.

Benfey, T. "Last Word on Polywater?" *Chemistry* 46(8) (1973):4.

Bennion, B. C., and Neuton, L. A. "The Epidemiology of Research on 'Anomalous Water.'" *Journal of the American Society for Information Science* 27 (1976):53—56.

Deryagin, B. V., and Churaev, N. V. "Nature of 'Anomalous Water'." *Nature* 244 (1973):430—431.

Gingold, M. P. "L'eau Anormale: Histoire d'un Artefact." *La Recherche* 5 (1974):390—393.

Hildebrand, J. H. " 'Polywater' Is Hard to Swallow." *Science* 168 (1970):1397.

Kamb, B. "Hydrogen Bond Stereochemistry and 'Anomalous Water'." *Science* 172 (1971):231—242.

Popper, K. *Unended Quest.* Glasgow: Fontana, 1976.

Prigogine, M., and Fripiat, J. J. "Anomalous Water: A Possible Explanation of its Formation and Nature." *Chemical Physics Letters* 12 (1971):107—109.

Schuller, D. "Zum Problem 'Polywasser'." *Naturwissenschaften* 60 (1973):145—151.

Taylor, E. "The Great Polywater Doodle." *Oak Ridge National Laboratory Review* 4 (1971):12–17.

Chapter 8

Allen, L. C., and Kollman, P. A. "Theoretical Evidence against the Existence of Polywater." *Nature* 233 (1971):550–551.

Auerbach, S. "Polywater: It Played Its Dirty Tricks for 10 Years." *New York Times,* September 2, 1973.

Davies, R. E., Rousseau, D. L., and Board, R. D. " 'Polywater': Evidence from Electron Spectroscopy for Chemical Analysis (ESCA) of a Complex Salt Mixture." *Science* 171 (1971):167–170.

Deryagin, B. V., and Churaev, N. V. "Anomalous Water." *Nature* 232 (1971):131.

Howell, B. F., and Lancaster, J. "Polywater: Is it Silica Sol?" *Chemical Communications* (1971):693–694.

Pethica, B. A., Thompson, W. K., and Pike, W. T. "Anomalous Water not Polywater." *Nature* 229 (1971):21–22.

Rousseau, D. L. " 'Polywater' and Sweat: Similarities Between the Infrared Spectra." *Science* 171 (1971):170–172.

Chapter 9

Bascom, W. D. "Polywater." *Journal of Physical Chemistry* 76 (1972):456–457.

Kuhn, T. S. *The Structure of Scientific Revolutions.* University of Chicago Press, 1970.

Larrabee, E. "Science and The Common Reader." *Commentary,* June 1966.

Madigosky, W. "Polywater or Sodium Acetate?" *Science* 172 (1971):264–265.

Salomon, J. J. "Science Policy and its Myth." *Public Policy* 20 (1972):1–33.

Chapter 10

Bernal, J. D. *The Social Function of Science.* New York: Macmillan, 1939.

Bikerman, J. J. "Anomalous Water." *Nature* 245 (1973):343.

Greenberg, D. S. *The Politics of American Science.* Baltimore: Penguin, 1969.

Koestler, A. *The Case of the Midwife Toad.* London: Hutchinson, 1971.

Langmuir, I. "Pathological Science." In General Electric Research and Development Center report 68-C-035, edited by R. N. Hall, pp. 1–13.

Maugh, T. H. "Soviet Science: A Wonder Water from Kazakhstan." *Science* 202 (1978):414.

Merton, R. K. *The Sociology of Science.* University of Chicago Press, 1973.

Partington, J. R. *A History of Chemistry.* London: Macmillan, 1962. Vol. 3, pp. 379–381.

Schaffer, S. "Polywater Research and Pathological Science." M.A. diss., Harvard University, 1976.

Wade, N. "Why Is Scientific Honour Giving So Sloppy?" *Trends in the Biochemical Sciences* 4 (1979):N186–N187.

Watson, J. D. *The Double Helix.* New York: Atheneum, 1968.

Ziman, J. M. "Some Pathologies of the Scientific Life." *Nature* 227 (1970):996.

Index